A Fenland Smallholding

A Fenland Smallholding

PAM BOWERS

THE CROWOOD PRESS

First published in 1986 by
THE CROWOOD PRESS
Crowood House, Ramsbury,
Marlborough, Wiltshire SN8 2HE

British Library Cataloguing in Publication Data

Bowers, Pam
A fenland smallholding.
1. Farm life - England - Fens
2. Farm life - England - East Anglia
3. Farms, Small - England - Fens
4. Farms, Small - England - East Anglia
I. Title

630′.92′4 S522.G7

ISBN 0-946284-37-7

Dedication

To Rick with whom I shared this experience, and to our children who make all we do worthwhile.

Acknowledgements

Line illustrations by Pete Westcott

My special thanks to Audrey Lloyd, the late Herbie Cousins, Eric Gosling and Jack Garratt; and everyone else who helped us so much along the way.

Parts of this book first appeared in serial form in *Home Farm*.

Typeset by Alacrity Phototypesetters, Banwell Castle,
Weston-super-Mare
Printed in Great Britain

Contents

PART I

CHAPTER 1

Small beginnings

John Seymour and his kind have told us how to do it, but we have never found a book written on what happened when someone did! This is the story of small beginnings; how a family started with nothing and worked up to seven acres. It tells of all the mistakes we made along the way; and rather than being a 'how-to-do-it', a more apt title would have been 'How not to be a smallholder'. I should explain here that 'we' at the stage of our lives at which this story begins, consists of Rick and Pam (both twenty-three years old) and our small son Jade, aged two.

We were living what could be termed a very normal sort of existence, in a semi on an estate (in Hertfordshire, fifteen miles from Cambridge) strung out on a mortgage. Many people have asked us why we moved out of the town - the exact thoughts have been lost in the passage of time, but it was mainly to get away from having to work in monotonous jobs to provide the income to fit in with the set in which we moved; from our seemingly pointless lives.

It was sometime in 1974 that we began to stand out as 'being a bit odd' to our neighbours and friends. We began to buy eggs from a local farm, instead of those off the supermarket shelf. We made our own bread from wholemeal flour. The bread strike came in the December of that year and we never bought bread again. Later we were to buy in an Atlas grain grinder and milled our own, buying wheat direct from the farm. It paid for itself in the first year, whole grain being a quarter of the price of its ground counterpart. We sprouted seeds on the kitchen window-sill in muslin covered jars - alfalfa and mung beans, professed to be higher in protein eaten that way. We grew herbs in yoghurt pots, and strawberries, purple sprouting broccoli and runner beans intermingled with roses and pansies in the flower borders. We were fortunate in that the garden already contained three half-standard apple trees, a few raspberry canes, blackberry brambles and a peach tree that had leaf curl and never yielded anything. Then we dug up the lawn and filled in the goldfish pond.

Having no gardening knowledge whatsoever, we just put everything in, all in the same manner, generally squashed together, as we were fast running out of space. At least a quarter of the garden area was taken up with a paved patio which was a waste, but could not be helped. We collected leafmould from the heath, shovelled up any horse manure found on the road and made compost from all our vegetable waste; all to add humus to the chalky soil. When we turned up to my sister's wedding, we sniffed ourselves anxiously, for we had been collecting bags of manure that morning, humping them in the back of the estate that was soon to carry everything from meal sacks and building materials to goats.

We decided to become vegetarians as a protest against factory farming methods. When friends came round to Christmas dinner, we served Chestnut Pate as the main course and a disappointing plum pudding, that tasted like bread pudding, for dessert. But the men still dozed off afterwards, just as though sated after turkey and all the trimmings. We bought our groceries from a health food shop, the nearest of which was fifteen miles away. There were not as many wholefood stores about then as there are nowadays.

Chestnut Pate

1 turnip (chopped)
8oz chestnuts
2 sticks celery (chopped)
2 large onions
2oz margarine
seasoning
8oz breadcrumbs
2 beaten eggs
1 tablespoon parsley
1 tablespoon lemon juice
2 tomatoes

Method Simmer turnip, celery and chestnuts for ¾ hour. Drain and mash. Sauté chopped onions in margarine. Combine all ingredients, with a little water if necessary, to make a dropping consistency. Bake in a lined bread tin at 350°F for 1½ hours. Serve with stuffing balls and lemon.

Stuffing Balls

1 lb breadcrumbs
8oz fat
4 tablespoons parsley
2 tablespoons mixed herbs
1 small onion (grated)
1 egg
grated rind of 1 lemon
1 tablespoon lemon juice

Method Combine all the ingredients. Form into balls. Bake for 20 minutes on a tray in a melted knob of fat.

We made wine from anything we could lay our hands on - rhubarb, dandelions, elderflowers, elderberries - gathered from the wastelands around the railway line; each one more gruesome than the first. We had never spent a great deal on drink, which is probably why we have never felt the need to perfect our method. Someday, when we find the time, we will get the proper equipment, rather than water ewers and anything else that comes to hand, and do it properly. We made ginger beer for the children by an excellent recipe, that does not require the complications of setting up a plant.

Ginger Beer

2 lemons
1oz bruised root ginger
1 lb sugar
¼oz cream of tartar
1 gallon boiling water
1 teaspoon yeast

Method Crush ginger, add sugar and cream of tartar. Pour on boiling water and stir well. Cut lemons into thin wafers and float these on the water. Cover with a cloth. When cool, sprinkle on the yeast. Leave for 2 days. Strain off yeast, ginger and lemon. Bottle in screw-top bottles. Leave for 3 days.

We used our spare time differently, often going for long country walks and always coming back with something useful we had found – pocketfuls of nuts perhaps or bundles of kindling. We often went up to Royston Heath on Sunday mornings and would come back with a boot-load of fallen and dead wood. Once the fuel had accumulated, we opened up the fireplace, setting aside the electric storage heaters. Some mornings I would walk with Jade up to the woodland on the far side of the park and come back trailing seasoned branches of trees.

We often picked up articles of clothing and other knick-knacks from jumble-sales, like the fur coat I unearthed for twenty-five pence, an adult's hand-knitted, Fair Isle cardigan for ten pence, a 'Biba' skirt for only a matter of pennies, and stopped frequenting the large chain stores. I also made a lot of our clothes, knitted or patchwork (in recycled material), which was very popular then. The latter I would often dye to soften the overall effect. I had time to mend and repair our clothes in a very thrifty fashion, until I was patching patches; and so Jade was dressed very well in an assortment of hand-made clothes, jumble oddments and clothes bought from small establishments producing garments which were individual creations. Jackie and Olaf Senff, some friends of ours locally, had a small business that turned out beautiful patchwork jackets and trousers in leather, hand-sewn by workers in their own homes, that sold for high prices in the best shops of Cambridge and London. We have just such a jacket. It has passed through each of our four children and is still as good today as it was eight years ago. Another firm started in a similar small way about that time, selling clothes in kit form – the first of its kind – and it has gone from strength to strength, opening several branches in large towns that hold prominent high street positions. We supported this establishment, because their goods were different, in unusual colours with unique prints.

Toys were bought along the same lines, avoiding plastic and seeking out locally made products. Friends who had spent an equivalent amount of money buying mass-produced toys for their daughter were amazed when we spent our £16 (I cannot imagine what it would cost today) on a fine, hand-made rocking horse. However, our purchase still stands, adding colour to our sitting-room. It has given pleasure to many children, who seek it out as soon as they walk into the house, and it will still be good for future grandchildren. We made Christmas and birthday presents when we could and from then on always produced our own cards. That first year, I remember they were arrangements of dried grasses on a sprayed gold background.

Then I came upon *Seed - Forum for Organic Living* one day - a magazine that was around at that time. It excited us greatly, seeming to echo all our ideals. When we went to the library, we came back with all the books we could find on wholefoods and herbs; among them, Juliette de Bairacli Levy's books, especially *Natural Rearing of Children* (although we did not use all the ideas in it). Friends of ours were distributors for her herbal products in the UK and we came to meet her daughter Luce, whose upbringing is described in the book. In fact, I helped out with the packaging of orders for a short period when the business was being transferred. Then we found *Fat of the Land* - John Seymour's own account of his family's early days of living a peasant-like existence, and not long after, a second book of his was published *Self-Sufficiency - The Art and Science of Preserving your own food.* I immediately put in a request for it at the library, and when it came I religiously read it from cover to cover, over and over again, and renewed it as many times as I dared!

We did all we could, little things like saving the top of the bottled milk for cream, but looking back we could have done much more. For instance, there was a huge communal stretch of grass in front of our row of houses - quite suitable for tethering a goat; and all those lawn-clippings that the neighbours filled their dustbins with were free for the asking. Rick could quite easily have put together some hutches on the patio to breed rabbits, and a few hens in a run would have been possible, even without a cockerel to wake the neighbourhood at dawn!

We went down to the West Country for a holiday and had a half-hearted look at property there through rose-tinted spectacles. But it was not until the approach of spring in 1975 that we realised we really wanted out! We had friends living in two different community situations near Royston, keeping a few animals, growing their own food and living close to the land; we had made a couple of attempts already at Field Crescent (as our house was called) at living with people, but both had failed miserably, bringing a strain on our relationships. We put it down to being too set in our ways - I hope that is not another term for being selfish – but communes were not for us.

It was all part of the hippy philosophy at the time, to drop out of the system. Most of the hippy exodus was making for Wales, to buy up a derelict and keep a goat. This is where we first tried (if that is what the brief sighting could be called). We drove down to Herefordshire on Friday night when Rick finished work and slept in a lay-by in the back of the estate. We did not seem to notice that it was cold and damp, only being mid-March, but the car did. It refused to start the next morning, and by the time we had received assistance from the nearest garage, the morning was well-advanced. We got to Ross-on-Wye and looked round the estate agents; they were all just shutting up and we realised it was half-day closing on Saturdays. We went on to Hay-on-Wye, where we had spent a beautiful few days once, and just looked at house-details through locked doors – the prices reflected the magnificent scenery!

After this laughable attempt, we went up to Lincolnshire one Saturday. Here was definitely a more realistic price range to suit our pockets. There were numerous cottages of character in the £3,000 bracket, with large gardens. We had already been made an offer for our property, and therefore knew that we had £7,000 + at our disposal. It was all very fortunate; we had bought just prior to the big property boom. We found ourselves winners at the other end too; the self-sufficiency craze had not caught on yet, and there was an absolute wealth of cheap property with small plots of land from which to pick and choose.

We knew absolutely nothing of the shortcomings of old houses – rising damp, penetrating damp, dry rot, wet rot, death watch beetle, structural faults – we were just looking for a cottage with roses round the door and land extending to the acre mark, be it silt, sand or heavy clay. We seriously believed however poor the soil, we had the capabilities of producing crops with 'Findhorn' dimensions by our organic methods. How naive we were! My father had suggested we wait a couple of years – think the matter over, save up – but, as always, we jumped in feet first and I am glad this time we did; otherwise it is doubtful whether we could ever have afforded the move. Many we meet spend so long weighing up the pros and cons of this way of life that they never actually get round to doing it.

The initial scouring of rural Lincolnshire was not fruitful. There was the ideal cottage in the district of New York, attractively situated by a windmill, but the purchase of additional land was out of the question to extend the limited area offered for sale with the dwelling. Another charming cottage with a magnificent cooking range and large grounds overlooked an horrific housing estate. A third on our list in the Old Leake Commonside area had the rural situation we craved but we found the house to be spoilt in its modernisation. Then we came across an advertisement in an *Exchange and Mart* for a two-bedroomed cottage with three commercial glasshouses and an acre of land, near Wisbech, Cambridgeshire, for £6,000. It seemed a dream come true, but on telephoning the advertiser we were told a couple were in line for inspection the following day. We automatically assumed that such a bargain would be snapped up at once, but we could not expel it from our minds. We continued house-hunting, but everything looked inconsequential in comparison and a few days later we rang again – 'Ivy Cottage', Walpole Marsh (as was its address), was still on the market. We arranged to view on the Saturday.

CHAPTER 2

Down Honeysuckle Way

Walpole Marsh is a fen village situated on the Cambridgeshire-Norfolk-Lincolnshire border. Once through the market town of Wisbech, we travelled what seemed miles into the very heart of the country (six in fact!). It was 19 April and cropping was in full progress. It was predominantly a market garden area and the land was broken up into small lots. Every crop imaginable seemed to be growing - from the wheat, barley and sugar beet of the larger farms through roses, orchards, soft fruit, cut flowers, cauliflowers, lettuces, dairy pasture to pony paddocks. The strawberries need a special mention; they were grown everywhere - in allotments, back gardens, market gardens and odd corners of farms. Everywhere was a hive of activity; I craned my neck - there was so much to take in.

Then we became lost, which was to be the story of our lives all the time we were there. The main problem was that the Marsh was only signposted from one direction, but we asked a native and thus tracked it down. Friends and family visiting for the first time could never find us; the septic tank emptying crew spent an afternoon driving round without success, only to return to the depot at dusk for a second attempt the next morning; our poor overworked GP wasted an hour travelling around, when I had been discharged from hospital with Dicken, our second born.

We fell in love with Ivy Cottage from a distance of 150 yards, from the top of the private lane down which it was situated. A horse was tethered at the corner - it was a real country community and we knew immediately that we wanted to be part of it. The shabbiness and peeling paintwork became imperceptible to our blinkered eyes; we registered only the climbing clematis around the door, the ivy growing up the white walls, the brooms either side of the gate and, above all, the roses, of which there was every kind imaginable. What a glorious array of colour they were to become as summer advanced!

At the back of the house, we were thrilled by the mass of

shrubbery (it was the sort of cottage garden you could hide in) - lilacs, japonica, forsythia - it took us months to discover them all. There was a magnificent honeysuckle plant that trailed up the coal store, spilling out and overflowing, completely obliterating the corrugated ugliness and still it crept on, forming a covered archway that led to the concealed hidden doorway, behind which stood the outside lavatory, shielded by a high privet hedge. Friends nicknamed this 'honeysuckle way'. Here also stood the yew tree, where the guardian spirit of the house lives (many stories tell of personal injury to fellers attacking this tree). But my favourite was the old-fashioned rose tucked away behind the currant bushes, with its soft hue and delicate perfume.

We were content to postpone viewing the house until last, although it was obvious now that Maud, the vendor, was doing her best to hold back the dilapidation of its interior for as long as possible! But it worked with us; we were absolutely transfixed with the land and all it had to offer. The glasshouses seemed huge; two fifty-six feet by twelve feet (one freshly set out with tomatoes, the other with climbing french beans), and one twenty-six feet by twelve feet (also tomatoes). The majority of the land (prime silt on sand we were told) was laid out to strawberries, gooseberries, perennial cut flowers and sweet williams, all except a tiny strip up the side, reserved for this year's sweet williams (being a biennial crop). There was also a block of onions and a few cabbages, potatoes and carrots. On the headland stood two plum trees and a pear. We did not notice that the outbuildings were in a state of woodwormed disrepair; all we took in was their number - two garages, a large barn, three other sheds and the coal store.

There was a conservatory on the back of the house, leading to a long, narrow kitchen. The appearance of the living-room was of organised clutter - too much furniture, of a 1950s' style, was crowded into its ten feet by ten feet dimensions. Ornaments and photographs filled the mantleshelf of the hideous tiled fireplace and any other spare surface space there happened to be. The furniture spilled out against the paned front door, blocking penetration of daylight and visitors, and a great deal of manoeuvring had to be undergone before entry could be gained for our inspection of the other sitting-room which was of identical proportions to this. Following on from here was a room of similar size to the kitchen, where a porcelain sink (later to become a herb garden) and a bath - not plumbed in - were housed. This room had access by ladder to an attic, and also a wide door with a gigantic key led to the back garden, in which was sited the cess pit and the soft water tank.

To gain admittance to the upper storey, we ducked through a door in the corner of the sitting-room and up a dangerously winding staircase. At the top we found ourselves in a spacious bedroom. The ceiling sloped dramatically down to the low window; one had to stoop double to look out, over allotments and the Marsh. This comprised a main road, where the chapel and the pub (all the amenities the village possessed) were situated, and a road leading off from either end. We passed out of what would be the master room into a second identical bedroom - the future nursery. The bedrooms had a southerly aspect, like the sitting-rooms below - with the kitchen and bathroom at the rear of the house, the north side. It was with much reluctance that we tore ourselves away from Ivy Cottage and went back to civilisation, to set the wheels in motion to make it legally our home.

The completion date was not until August; we therefore had four months to put to good use, to make plans. We booked a two-week holiday in Holland and looked at it more in terms of our last holiday ever! I enrolled for a mixed craft evening class, producing objects I thought would have a purpose in our new life - a surrey-stitched rug with Mexican motif, baskets, canework plant pot holders and a macrame doorhanging.

At the first signs of spring, we had planted our seeds. They were sown in a rare hodgepodge of containers - seed trays, plant pots, tomato trays from the greengrocers, yoghurt pots, margarine tubs, anything scroungable in fact. We put them onto the patio during the day and at night they were brought back indoors, out of danger from frosts, so carefully did we nurture them. It was a real performance. But, as soon as the conveyancing was going well, we piled all this paraphernalia into the back of the car one Saturday and headed for Walpole. The section of bare land allocated was only small, but into that patch we crammed all our vegetables - cauliflowers and other brassicas no more than nine inches apart; sweet corn similarly squashed; potatoes too, transplanted from our garden in full leaf. We did all this and everything thrived at our amateur touch. To this very day we have never had such fine crops - maybe it was beginner's luck. From week to week we did not see our vegetables to tend and weed them, and although it was an extremely dry summer and we could do no watering whatsoever, it all grew well - even the potatoes!

We took photographs to keep the place alive in our minds, even the camera noticed the damp patch on the gable-end wall in our bedroom, that our eyes failed to detect - surely a bit ominous when there had not been any real rain to speak of for months! We needed the least excuse to visit Ivy Cottage and the chance came at Whitsun. Maud wanted to discuss prices with us for the dead stock we had agreed to purchase - articles such as a Rotavator, greenhouse heaters and an assortment of gardening tools. We took this opportunity to redecorate the outside of the cottage at the same time, in readiness for moving in.

We sold all our modern furniture - the white chipboard settee suite and armchairs, the pine dining suite, with imitation Welsh dresser. Then we searched the antique and junk shops for some traditional furniture - two leather-seated dark wood chairs, half a dozen pine chairs with a solid pine table to match, but minus the drawer. Alec, in whose shop we discovered it, promised to put one together in such a skilful fashion the difference would be indistinguishable - he did, eventually. It was delivered at 11 p.m. the night before we were due to move!

We made a dramatic gesture at discarding all the mechanical equipment we possessed, associated with the industrialised society - we sought a peasant existence, to live better on less. The vacuum cleaner went, followed closely by the television and then the washing machine. But the latter was only shortlived as I realised that I preferred to spend the hours a washing machine saved me doing other things. I became grumpy stuck indoors - especially as the pile of dirty clothes got higher as our family grew. A day confined to the housework by the weather was a day wasted.

Then one day we received a message to call our solicitor. There had been a change of circumstances, and the completion date had been brought forward - 29 June was the magic date. This season's crops still belonged to Maud by the terms of the contract, the only difference being that we were to give her reasonable access to harvest them, while we lived in the cottage and she lodged with her friend. We ordered the removal van and did all the last minute things that were necessary (including buying a stock of bulk wholefoods, brown rice, dried fruit and so on, that we would be hard pushed to produce ourselves in this climate); it was as if we were journeying into the wild unknown, not just sixty miles away into East Anglia! And so this is how we came to be escaping from the neighbours' son, playing music into the early hours; how we came to be leaving behind our red, yellow and green painted sitting-room, with the square foot of bare wall that was revealed when the picture was taken down. The message in its place explained all ... 'sorry - ran out of paint'!

CHAPTER 3

Where the wind blows

So we came to the Marsh in the midst of a glorious summer, our initiation into the Fens, where the wind blows three hundred days of the year. After refreshment at the village local, the removal men swiftly unpacked our goods and chattels, ranging from my six foot trifid-type rubber plant, which promptly curled up and died in the dramatic change of environmental conditions, to the enormous collection we had built up of 'things that would come in handy one day'. What did not go through the front door, had to go through the bedroom window, which had been removed in readiness; the narrow winding staircase being impossible to negotiate with double beds and cupboards. Somehow it was all achieved without too much mishap and later, when everyone had left and we retired to bed, it seemed as though we were the only people alive. I will never forget the complete and utter stillness of that first night.

In those early days, the sun always shone in every sense of the word. We found much to do (although it was done in a pottering fashion, we did not know what hard work was yet), and there were always new things to come upon and to learn. The tradesmen made their introductions, but we declined their services – 'No', we said to the butcher, 'We are vegetarians'. 'No', to the baker, 'We bake our own bread'; and from the newsagent, we only ordered the local paper, published once weekly.

We had already become acquainted with our next-door neighbours from Rosedene, when we were still travelling back and forth decorating. Audrey had gone out of her way to welcome us with her cups of tea and biscuits (which were to be enriched with brandy on cold days). She had the time to chat on ... about her elderly mother, known to one and all as Granny, about Herb (he did the garden) now one of the family since his wife died, about her two daughters – Marie, in London with her family and 'young' Audrey in her thirties living at home and courting a farmer on the 'other side of the river' (Tydd Gote, eight miles away). After a few minutes in their company, we felt we had been part of their family all our lives!

None of the locals could make us out. Maud had spotted the guitars being unloaded and rumour quickly had it that we were a pop group. With Rick's long hair and our mode of dress (which would have blended easily in city surroundings, but stood out conspicuously here) I suppose this was understandable. You must remember this was 1975 and we were the first of our kind to invade the area. To keep a nanny goat or two was as yet unheard of and considered most odd. Wherever we went, people knew who we were - such was the village grapevine and the essence of country life.

We went to that centre of the local community - the pub. It consisted of one spartan room, the former sitting-room of the private residence, where bench seats and armchairs intermingled and the bar was sited in an adjoining room. Thus the youngsters were at liberty to guzzle coca-cola amidst the ribald laughter of the revelling men, content in the knowledge they would be plied with crisps by the hospitable landlady, Kath. On this our first visit, the men questioned and quizzed us - 'Were we artists, musicians?' They could not understand it at all.

It was over a drink that we first made the acquaintance of Jack and Gogga, farm-workers whose allotments Ivy Cottage overlooked; and of Greg (shortly to become a divorcee with custody of his two children) who kept a bit of derelict land and a semi-derelict pig-enterprise! There was his father who lived in one of the council houses and was partially sighted, but rode a bicycle in a wobbly fashion. Whether it was luck that got him around safely or the fact that everyone was aware of the condition of his eyes and made allowances, I do not know. There was poor overworked Kath herself, who not only managed the pub by night and worked by day packing sterilised soil for the small compost manufacturing concern on our western boundary, but also looked after her family, an invalid husband and a patch of land. Everyone possessed at least a few square feet of the earth's surface and made what they could out of it.

It was quite a usual sight to see Eliah draw up in his polished racing trap of a fine Sunday morning, to leave his pony grazing contentedly on the roadside, while he span one of his tall yarns or pulled his harmonica from his pocket, as he downed a pint in the comfort of the old tavern, the same today as it must have been a century back. Eliah was a Romany who had put roots down in the Walpole villages; buying a field and building brick amenities alongside the vans of his large community, all seemingly related. Other vans would come and go, drawn like a magnet to the camp when there was land work to be had; the news spreading without the use of telephone or the Royal Mail.

But that slow pattern of life we had settled into so easily was not to be. We came down to earth with a crunch, exactly one week later. Little tell-tale signs had begun to manifest themselves, like the wallpaper that was literally hanging off the walls in the nursery, which had only been decorated two months previously (bearing in mind too that this was the great drought); there was the wardrobe which turned out to be of blue asbestos. The dire necessity was to protect our clothes, which were rapidly displaying signs of mildew. There was the linoleum tacked to the kitchen walls concealed beneath wallpaper and the hardboard to achieve the same end in the sitting-room. We began not to venture into the bathroom unless it was absolutely essential (such as for access to the tap through its exterior door to water our thirsty crops) because the putrifying smell of damp was nauseating and the asbestos nailed to the walls in there was obviously a last resort! The whole surface of the ground floor rooms was concealed in layer after layer of linoleum. Peeking cautiously beneath, we found tiles on earth – the cottage had no foundations.

We chanced on an advertisement in the local paper of someone offering free estimates for damp-proofing and thought we might as well enquire, and the nightmare began. The chap turned up with a device that detected rising damp – it buzzed ominously when encountering any moisture and then a reading could be taken. The instrument took off on our premises like an automaton from outer space – 'beep beep beep beep beep'. It shook and quivered happily in robot fashion. Damp was revealed not only on the downstairs walls, but half-way up the bedroom one too. The house, formerly two tiny cottages, had been built two hundred years ago, when the men used to build the walls only as high as they could reach. Then they would construct the floor on top and use this to reach up to complete the upper storey. Hence the ceilings of Ivy Cottage were six feet high – five feet ten in parts. The gentleman looked grim – this was a very serious case indeed. He forecast what dreadful diseases a child in these conditions could contract. The whole thing seemed far bigger than he could handle. He suggested that we contact the local authority for an improvement grant and then beat a hasty retreat. 'Incidentally, the ivy is eating into the brickwork at an alarming rate', were his parting words of comfort.

The first step we took was to cancel our holiday in Holland, arranged for August. We would need all the money we could get. The second was to rip down the ivy, before it could cause further destruction, and so Ivy Cottage became ivyless and consequently even harder for strangers to track down! Then followed months of hindrances, hitches, red tape and frustration. The Environmental Health Officer came out from the council to inspect the property, to give his assurance that it would be good for another thirty years after improvements had been carried out. He explained what the grant constituted and about the five year clause. The condition was we had to sign an undertaking that it would be our home for this period of time; failing which, the grant plus interest had to be paid back in full. To qualify for the grant (which at that time was fifty per cent of the repairs needed, up to a maximum of £3,200) we learned that far more had to be done than just the damp-proof course we had at first envisaged - the list included:

Rewiring
New roof
Renewal of rotten and woodwormed timbers (in other words all the timber!)
New staircase to come up between the two bedrooms with dividing landing space
Renew all perished plaster

In short, just the walls and window frames were to remain.

It was now in our hands to obtain estimates and have plans drawn up. One of Herb's sons, who lived on King John Bank (King John is said to have lost his jewels somewhere in the area - they have never been recovered), happened to be a builder of good repute, and in early June 1975 we approached him for a quotation to do the work. Week after week we called by, but no amount of bullying could hasten him. We found his eventual craftsmanship certainly to be of the quality reports had indicated. He was a perfectionist of the highest order and had he worked harder he could have been a very rich man indeed. But Denny did not seek riches; he was happy to remain a country builder, keeping a steady pace, working at that which he enjoyed most and being able to stop for a chat now and then or a 'cuppa' when he wanted. It was Christmas before his estimate was through; then it was the architect's turn. There was a fellow on the District Council, who drew up plans in his spare time and for £30 produced ours. The Council met once a month and on 2 February we heard that our submitted plans had been passed. We had been awarded a full grant of £1,560.73 representing fifty per cent of the total estimated cost, the work to be carried out within eighteen months of that date. Where we would get our share of the expenses from we did not know, but I cannot remember worrying unduly.

Denny arrived on 15 March to commence work – but we are moving too fast. There are the eight months initiation into our peasant existence we are overlooking – a winter (fortunately an unusually mild one) spent in extremely primitive conditions; of stand-up washes at the kitchen sink – baths were out of the question in that blustery hole with other horrors lurking in the dark unknown; of exits down honeysuckle way – a trip to the toilet became a social event when we had visitors, the shadows do not look half so sinister when you have a bit of company! There was one particular night when a violent storm blew down a tree making the lavatory inaccessible! I think the best thing about our complete lack of amenities was the £11 quarterly electricity bills.

We came to love that continual wind – on hot days it was the breeze that made it bearable to work, and I really revelled in the gales; the feeling of the force of nature that is all powerful; the complete abandon with which one had to struggle to carry out the simple everyday tasks of carting fodder out to the animals, using all the strength one possessed. The darker side of it all was lying in bed at night, listening to the storm raging about you, wondering how many of the dutch lights in the greenhouse would be broken this time; the first £15 in each case being payable by us the insuree!

Every single day of that first winter we worked on the land. It seldom rained; all the fuel we used was broken Holland trays, (cleared up from around the place) and we only lit the fire when we stopped work at twilight. We sincerely believed this was what winters were like, being unaware of the seasons in a town, not realising it was the build-up to the eighteen month drought. Should we wish to finish earlier than the usual 4.30 p.m., we would 'call it Friday', whatever day of the week it happened to be. It was traditional in those parts, being an agricultural area, to work a shorter day on Friday, the last day of the week. Audrey would pass by the hedge, as she and Herb bustled round clearing away tools, scales, trolley and boxes for the week and shout out: 'We're calling it Friday'. We came to know them better and better. They watched our efforts at doing things they had been brought up to with amusement, particularly Rick's initial attempt at operating the Merrytiller. As he took off in a cloud of dust, not realising the throttle had to be adjusted, our neighbours roared with laughter. Though we did have our fair share of laughs at their expense too!

Granny in her nineties was extremely proud of her great age, every time she reached a birthday, instead of stating 'I am now ninety-five' or whatever, she would say 'I am in my ninety-sixth year'. She lived with Audrey, who jumped to her commands - although the old lady herself 'cooked them a nice dinner every day' as she never tired of telling us - but it was her son, Bob, who took her out on Sundays who was the apple of her eye. Any acquaintance that gave birth to a baby of the fairer sex she pitied.

Down the lane lived Aunt Ida, a prim spinster who had once had a fleeting relationship with a young soldier she met on a train. Aunt Ida had passed her mid-eighties and therefore was considered a rival by Granny, made worse by tinges of jealousy, because Audrey had elected herself responsible for the old lady, being her only living relative in the locality albeit on her husband's side. Rick once saw Granny scuttle away, chuckling like a schoolgirl, as Ida approached with her knitting bag, indicative of an afternoon to be spent in their company at Rosedene. On other occasions, he caught sight of Granny spying round the corner of the house, checking that Herb was not slacking during Audrey's absence in town. Yes, she was a lovely old lady. Alongside Aunt Ida lived Mrs Playford, another pensioner. Jade could not pass them by without having sweets pressed into his hands - they really spoilt him. It was all a nice part of country living that we knew we could not walk up the lane without stopping to pass the time of day with one or other of the oldest inhabitants.

We did all the things we had moved into the country to do, with such serious dedication that it amounted to near fanaticism; we bought the chickens and the goat; we worked the land and reaped the harvest, storing it as the housewives of bygone days always did - in salt, pickles, jams and bottles. It was the kilner jars we found to be expensive beyond belief and we went into a tiny old-fashioned hardware store down Norfolk Street in Wisbech, that sold plain, simple goods, where we haggled the price for buying the jars in quantity. We went fruit picking to pay the bills – a fact of life we had so blindly overlooked in our quest for the 'good life'. But I will tell you about all this in detail in the chapters that are to follow. Looking back they were happy times indeed and we are now able to laugh even at our mishaps and disasters!

The enthusiasm coupled with supreme naivety is reflected in a letter written to friends that Christmas:

Ivy Cottage
3 December 1975

Dear Andy and Angie,

Christmas always seems the time to be in touch with everyone we should have written to long before.

We've been here five months now – just beginning our first winter. The house needs attention; it's very damp and cold. We're in a very exposed position and the wind always blows.

We are slowly getting the land together and are now at least self-sufficient in vegetables. We've bottled and dried lots of fruit and done jams, pickles, salted beans etc. Rick's been busy with gallons of cider and we've tried several wines.

We've a dozen hens, who keep us supplied in eggs and a cockerel, Barnaby, to keep them happy. He is very 'hen-pecked' and never cock-a-doodle dos! There's a kid goat in the stable and until this weekend we had one in milk on loan. However this spring will see the arrival of a newly-kidded nanny, who should produce approximately a gallon of milk per day. Then we will make 'Norfolk' cheese and butter.

We did fruit picking during the summer (blackberries, apples, plums, pears) to pay the bills, but next year we hope to do this from our soft fruit and tomato sales.

The car is our main extravagance, but unfortunately transport here is a necessity, so we must wait until we have more land for 'the horse'. We bought £12 wheat which will keep us in bread, cakes, etc. for the year, but hope to grow our own sometime.

The goats we mostly tether on the roadside. We've put some land down to hay for them, but still have to buy oats at the moment. In the spring, we will plant potatoes and sunflowers for the hens. We are full of ideas – a duck pond on the lawn; a pig and even maybe a sheep for wool. Gradually things take shape.

Please come and see us if ever you can. We've plenty of room (if you don't mind the cold!). Well, happiness to you at Christmas,

Love Pam, Rick and Jade.

CHAPTER 4

Friends and neighbours

The 'demolition squad' moved in that March and pulled the roof asunder; with nothing but a tarpaulin between us and the heavens above, it was most fortunate indeed that this period marked the great drought of our history. However, the blazing sun that beat down on us day after day without respite brought problems of its own. The electrical system was in a state of limbo for a long time, being at a mid-way stage between disconnection of the primitive circuit, installed in 1950 when the 'electrics' first came to Ivy Cottage, and rewiring. For this reason, the refrigerator and washing machine were out of action. Our third and last modern appliance, the electric cooker, stood in the centre of the living-room and doubled as a water heater and food safe when the oven was not in use, offering some protection against the unremitting dust. Milk went sour within hours; margarine melted and went off rapidly and the mere sight of all this rancour did nothing for my morning sickness – what better way to celebrate our new life than with a baby? Facilities just did not exist for the laundry, be it hand-washing or machine, so whenever the clothes situation became drastic, we would pile it all into refuse bags and drive the ten miles to the nearest laundrette at King's Lynn. Not that we had to bother queuing for the driers, a few minutes on the line back home saw it dried and aired too!

Once the timbers had been renewed and the new roof was in place, we played a game with the builders or they with us, moving our few possessions from room to room. As Denny and Keith worked on one room, so we carted our chattels to the next. We had pared our belongings down to the absolute bare essentials, and stowed the remainder – ornaments, pictures, carpets, curtains, things that would only add to the clutter and soak up the filth – away in the garage to be attacked by mildew and mice. This is how we came to permanently reduce the frills of our home life somewhat. So we lived amongst builders' tools and rubble, with unplastered walls and the exposed bare rafters above us as we lay on our mattresses at night. The only housework was that done at the end of each day, as the workmen superficially swept away the surplus waste from their labours, and everywhere was thick in unpreventable grime. Generally we coped, but once in a while it all became too much – like the time I took to my bed with tonsillitis and the damp-proof coursing crew chose that very day to arrive with their pneumatic drills, that shook the very being of the house and my sick soul into the bargain! Another time, visitors turned up unexpectedly, friends we had not seen for a long time, eager to see what this 'good life' they had heard so much of, was all about, and that day we could not even make them a cup of tea. In those circumstances, with the facilities for cooking just about as inconvenient as they could be, we started relying on convenience foods again – even stopping off at the fish and chip shop after a hard day's fruit picking. The staircase joined the other rotten timber on the huge stack of rubble, hardboard and disintegrating linoleum in the yard. David, Audrey's young man, came over on his tractor for a couple of loads of hardcore for a farm track, but the majority stayed littering our access road for years, until we finally hired a skip when we were tidying up the place in readiness to move – the only time we ever had the incentive to really get the place looking shipshape. Eventually one room was complete, with electricity laid on and even though it was only a bedroom, we once again took up our bed and we made this our base.

We had been disappointed when the old doors had been condemned and so commissioned the new ones to be kept in character. We were thrilled with the final results – heavy tongued and grooved doors in dowelled pine with black Suffolk latches and hinges. At the front was a split stable-door fashioned in the same style. The doors were in keeping with the wooden floorboards upstairs and the pine staircase.

Although Denny's progress was infuriatingly slow on occasions, there was nothing we could criticise; all work done under his name was that of a first-rate craftsman, but not so that which he subcontracted out. The chap who came to measure up for the new staircase had to return three times before he got it right. We went to choose the bathroom suite, picking out a turquoise one as white is not exactly ideal with children around; but we found leak after leak when the plumbers were gone and we had trouble with the kitchen sink to the day we moved. The prospective buyer asked for a drink of water and stood watching whilst I grappled with stopcocks to provide such a supposedly simple request! The council representative carried out his final check and declared the cottage satisfactory on the provision of a storm porch, as a safeguard for the extra weight, on the bulging south-facing wall. We always thought it detracted from the cottage's plain appearance.

With the income from that year's short strawberry season and by working hard at fruit picking, we were at last able to pay off all our debts to Denny, who, with his workmen, walked out of our lives (bar the occasions the glasshouses needed repairing) and we missed having them always about, with their strong Norfolk accents and their country ways.

Whilst all this was in progress, we were living through one of the hottest summers on record. Sue, Rick's sister, came up to stay and we got into the habit of taking a midday siesta as it was just too unbearable to work at the height of the sun, and so everyone went to the river bank, the coolest place around. There was Herb and Andrew (Audrey's grandson) and Rick and Jade, all cluttered with fishing equipment; and the women with the picnic lunch and books to read, but generally the heat overcame us in a very short time and the murmur of voices would fade as we dozed. Sometimes an eel would be caught to be skinned and gutted later, more as a gesture to demonstrate to Jade the satisfaction of catching your own supper, than for the bite of protein it yielded in proportion to all the washing-up involved!

If we could muster the energy, we would head for the marshes, on the other side of Sutton Bridge, buying ices on the way. It was sheer delight to amble among the mud-flats by the lighthouse, maybe collecting samphire or sea-lavender, or some driftwood. It was on one of these occasions in 1976, as we walked down to the sands, that we came upon millions of dead ladybirds. It was impossible to tread without crushing the carpet of shells.

It was in the early morning, around 5.30 a.m. when perhaps you might be lucky enough to feel a light dew, that we ran the holding and did what needed doing. We would plod on weeding and harvesting with the sun burning down until, near the point of collapse, we surrendered to the heat of the day, resuming later and working on until sundown.

The following year could not have been more different: it rained and rained. It was the Queen's Silver Jubilee in June and Rick's grandmother came to visit, invited for a week that spread to nearly three. During all her time with us, it was cold enough to run a fire. Hours were spent in torrential rain, picking flowers; when flowers are ready, off they must come without more ado, but to compensate we have never received such a good price for them as that Jubilee year. The people of the Walpole villages really did themselves proud in organising a programme of events, culminating in a children's party on the Saturday, with a barbecue and dance in the evening. This marvellous community spirit was to develop into other things. The Jubilee Committee, as they were called, brought out a newsletter, published monthly, that was to be an arena for the airing of views and news of local happenings, and a forum for advertising. It became obvious that an active village should have its own village hall and when we left, fund-raising efforts for one were still frantically going on. One of the biggest events each year was the craft fair, that the village held in November in King's Lynn (not having a hall of its own) where we bought our Christmas presents and plundered ideas!

Another outcome of the Jubilee celebrations which affected us more personally, was the contacts we made - up until then our social life there had been pretty limited. Now, we suddenly found that we had a whole new circle of friends in our own age group. But, although we shared common interests with our new friends and were held together by a bond - that of each trying to improve the quality of our lives - it was a full three years, in the summer of 1978, before we met others interested in self-sufficiency. We were at the annual local Goatkeepers' Society show, when we spotted an advertisement in their newsletter from someone wanting to share transport costs to the Practical Self-Sufficiency Open Day at Saffron Walden, and this is how we came upon John and Marilyn. It made a world of difference to discover like-minds, not more than three miles distant from our homestead. It gave us someone with whom to exchange views, sort out problems and discuss new projects; someone to take a general interest in what we were doing.

Then, one day, a friend arrived out of the blue and stayed a while. She produced out of her vast bag an address jotted down from an advertisement in a Cambridge wholefood shop, for local self-sufficiency contacts, leading to the possibility of forming a local group. We dropped them a line and in due course heard from Roy and Janet. Apparently they had been putting feelers out in all directions – libraries, wholefood stores, periodicals – and a gathering had been arranged at their own holding. It was a small cottage, that shook whenever a train went over the adjacent railway crossing, with an acre of land across the road, and at that first meeting the walls fairly bulged. This interest in self-sufficiency seemed to have no bounds, it spanned all age groups and encompassed a wide range of opinions and motives, including politics, ecology, alternative technology and those like us, who wanted to simplify their lives, virtually stepping back a century in time, transferring to a peasant economy, but with the advantage of advances in modern technology, from which to pick and choose knowledge should we wish. But we all had in common our obsession with a wholefood diet, keeping livestock and growing our own vegetables.

The group was a very loose organisation, with the rather grand title 'North Cambridgeshire and North West Norfolk Self-Sufficiency Group'. Roy sent out a news-sheet quite regularly, an opportunity to advertise produce, but by far the most important aspect that arose out of it was the bringing together of like-minded souls at the once monthly get-togethers in the comfortable informality of each other's homes. Once or twice there were speakers, but generally there was no set programme of events, and it was up to the individual to make what he wanted from it. Several of us organised a job sharing scheme – a day's work for a day's work; a chance to make boring tasks fun and to get jobs done that could use an extra pair of hands, haymaking for example.

The morris dancers came and brightened the streets outside the local pub; the open sewage beds froze on the Marsh and, near where Rick would mow grass for the goats on finer days, folk from the surrounding villages skated; Clyde, our third son, caused a stir by arriving when the village was cut off by a blizzard; young Audrey was finally joined in wedlock to David Fox and in due course produced an heir, young David, with a beautiful daughter, Charlotte, two years later.

Although our socialising enriched our days, while partying and jollification took place in the background, our daily smallholder's existence went on in the fore. We at last acquired a half acre allotment after years of applications, only to find it covered with every weed imaginable. But life was good.

CHAPTER 5
The fruit will follow

At Ivy Cottage, the year revolved around the strawberry season. When strawberries came, we had money and, just prior to the strawberry season (May/June time), we had none. In the strawberry season, we were frantically busy; often rising at 5.30 a.m. to pick some while the children were still asleep and the dew was on the leaves, and not sitting down until 9 p.m. for a milk drink before flopping painfully exhausted into bed for a few hours' deep sleep before it all began again. It was because of this pressure of work that we tried to organise the routine tasks as best we could, cutting them down to the minimum necessary. Cooking is an example - a few savoury flans and pizzas baked in advance so that our meals were as convenient as possible. The mending would be got up to date, stocks of animal food bought in and the house given a good spring clean; all in readiness for the strawberry season. At Walpole, we only picked fruit Monday to Friday, when transport ran to the market, so weekends at least were a chance to do the housework, go into town for the groceries and get on with the most pressing of the jobs on the land.

The season ran from about 20 June until towards the end of July - six weeks. During the first week, when the fruit were few and picked pink, the price received was high, and it was usual to send in quarter to half pound punnets to make the best price. Then followed four hectic weeks of hard picking when the full punnets almost filled the row; lastly, a final look for the few late fruit, when the price might have risen a little, as there was less fruit being sent to the market. This is all very approximate; a long hot spell of weather could bring the strawberries on all at once - a two-week freak season. A cold summer could mean slow, relaxed picking in dribs and drabs.

However, when the strawberries had been 'shacked' over, one could relax a little, get the land cleared up and start to enjoy the summer - a couple of trips to the seaside maybe, before we began fruit picking at Walpole St Peter, in late August, as we did each year.

Those first innocent days at Ivy Cottage, the sun always shone. We moved in at the end of June – the cottage now our territory, but with Maud still reaping its harvest. The strawberry season was in full swing and we watched her haste and frenzy – we ourselves settled into the pace of country life one imagines to exist from poetry and artists' impressions, awaiting the time when we should have complete possession of all.

Over our morning cup of tea, Maud taught us the basics of our new trade and how the produce was marketed. A family firm of wholesalers at Smithfield Market, Manchester, ran an open-top lorry to collect and transport the fruit to market during the strawberry season. The trays were picked up about 7.30 p.m. on weekdays only – and sometimes quite a bit later, as we were soon to learn when we stood waiting to hear of the previous day's auction prices; but it was an enforced opportunity to slow down, to exchange news with one's neighbours, who were all involved in the same activity. When the lorry did not run, early strawberries or any other produce could be sent up in the enclosed van of a local haulage company, but it had to be ready by 1 p.m. and did not travel as well, without the air circulating around it. We often met one ancient fen-man on the road, who had risen with the dawn to pull his fruit, cycling the six miles into Wisbech auction market with trays stacked on the bicycle trailer behind him. Maud explained that the wholesale firm supplied the punnets and trays and that the weekly cheque would already have all costs deducted – agent's commission, packaging, transport, porterage at the market. We chose to continue with these convenient arrangements and always found the firm reasonable and fair.

The summer was not all halcyon days however; there would be days of persistent rain, when picking still had to go on, and one summer while we were there, it did little else but rain. Dicken was a baby and thankfully a very contented one, as he had to be sat for the majority of the season in his chair, bribed with strawberries and anything I could think of to amuse him, in one of the greenhouses, from which position we were in view, engaged on any one block.

We gained experience, picking up tips along the way, such as topping up the punnets with the largest, most temptingly mouth-watering strawberries. Soon we found it was possible for me to handle our amount of strawberries (quarter of an acre maybe)

Our land at Ivy Cottage

alone, so Rick earned money by helping Audrey to pick and also to take the morning's strawberries to the jam-maker at Walpole St Peter. This fruit, not of such a high quality as that punnetted and already plugged – that is the stalk removed during picking – was weighed into stone trays. Prior to the commencement of the season, the buyer would call and give the grower a contract. During the time we were there, the jamming price remained constantly below market value, so it was only that first year, when we still had Maud's old plants, some being five years old and therefore of a poorer quality, that we bothered to obtain a contract.

Strawberries had been an untold luxury when we lived in the town. We would buy a punnet or maybe two during the season and dip them in a pile of sugar at the edge of the bowl. Now they were a matter of course. We popped them into our mouths as we picked, until we were sick of the sight of them. We dropped them into our breakfast dish of muesli and into our supper dish too. I piled more and more rejects into a gigantic freezer bag until the pace slackened enough to boil them into jam and smaller bags to use for fillings of cream cakes on winter days.

Once fruiting was over, the men whetted their scythes and mowed off the tops. When dried, the straw and leaves were burnt, thus returning valuable potash to the soil. The plants were also fed with special strawberry granules, applied similarly to broadcasting corn, either during the autumn or early spring. A favourite saying in Walpole was: 'If you've got the top, the fruit will follow.'

Strawberry plants are profitable for three years, after which they are ploughed in; so each year we replaced one block of Maud's. I collected our order of Cambridge 422s, during the planting period, September-March, from Fred Greenwood down the road, who sold certified stock (that is virus free) at £20 per thousand. I glanced admiringly around his six-acre holding, as he chatted. Theirs is an excellent example of an efficiently-run, small-scale enterprise, producing mainly salad crops (lettuces, spring onions, radishes), some strawberry plants and brassica plants. The place was always immaculate and each weekday they ran quantities of produce into Wisbech auction market. Lifting the plants, tied with binder twine into bunches of fifty, onto the wheelbarrow, I covered them with sacking to prevent the roots drying out. Dug plants will stay fresh, covered on an earthen floor for a fortnight or more.

Rick had marked the rows out, three feet apart, with a marker he had put together. With him at work now, I commenced with a small hand spade on what looked like a very long task indeed, until Herb and Audrey hurried through the gap in the hedge; Herb with hessian sack fastened around his waist by binder twine to insulate him against the cold winds (which was accompanied by fertiliser bag clad knees if the job called for it) and Audrey with dress and coat over trousers - her winter working apparel. Audrey showed me how to measure the necessary twelve inch distance apart, by dropping a plant either side as you walk down the grooved-out row. Herb followed explaining how to dig a wide slit with the hand spade, shaking the roots of the plant down into it before slipping the spade out, and stamping over the plant, to firm it into the soil, as with brassicas. The crown must be above the earth, but the roots loose and free, and safe against the frost. Before Rick's return that evening, the job was complete - thanks to our good neighbours.

Newly planted strawberries are called 'maidens'. After a year, they become first year pull, then second and third year pull, and afterwards ploughed in. Plants are only taken from maidens, the following autumn through to March. Once, we tried digging our own plants for sale; several days spent cutting, sorting and bunching earned us under £5 for 1,000 uncertified plants at

auction. In future, we just replenished our stock from the previous year's plants, setting them out into a well-manured patch where they thrived. Then finally the alleyways are sloughed – that is, the soil turned over, covering any grasses, weeds or strawberry plants encroaching onto the row. This operation is traditionally carried out with a Martin's spade, still only manufactured by the blacksmith in Wisbech who invented it.

In the spring, as the plants grow, so too do the weeds and much time is spent removing them. Then as the flowers form the crop is treated, as a precautionary measure, with a fungicide against rot and an insecticide, primarily against red spider, which curls the leaves and shrivels the fruit. A little time then remains to do some odd jobs at a reasonable pace before the first few strawberries are ready. The blackbirds seem to help themselves to a fair proportion of these, but there is no need to turn them to ripen like a new acquaintance did, on moving into the country from the Medway.

CHAPTER 6
Windfalls

We spent every summer on the Marsh fruit picking in the next village. I have fond memories of those halcyon days: the wonderful community spirit; the atmosphere; the hum of voices in the distance; all totally involved in the same thing, getting the harvest home. The work commenced with blackberries in mid-August and went through plums, dessert apples, pears and cooking apples to Russets and Ida Reds around Guy Fawkes' Night. It was a guaranteed income for us each year, but we earned every penny of it. It was piece-work, so much per pound for all blackberries (which went to the canning factory), by the box for other fruit destined for market and by the crate and half crate for apples and pears going into cold storage.

The workers would start arriving about 8 a.m. - local women dragging their offspring behind them and vast families of van dwellers coming back from their breakfast for the second stint of the day. The youngsters would begin the day full of boastful intentions to pick more than they ever had before, but very quickly the enthusiasm would be gone and one by one they would scuttle off to play. All except the gypsy children that is; they were expected to work alongside their parents. There was Benny with his wife and five dark-skinned offspring of stunning beauty. When he set down one full chip of blackberries, seven full chips went down. Each daughter was a dignified miniature of her mother; her trousers covered by a calf-length dress. On attaining adolescence, her long jet black hair would be unplaited and put up on her head - a kind of symbol of her advancing womanhood. It would then be covered by a brightly coloured scarf.

Long before the church clock chimed twelve, the children would return, with their purple-stained faces and bulging pockets of under-ripe apples, flocking round complaining of hunger pains. The weary women would yell at the little angels to 'push off' until lunchtime, cramming their grubby hands with crisps and sweets to pacify them. Then the tractors chugged into life as the full-time workers (who weighed and boxed the fruit and carted it to the yard) went off for their lunch. Everyone took this as their signal to collect flasks and sandwich boxes and seek out an upturned box or a tree trunk against which to lean, gathering together in small groups or family circles.

Foreign students of all nationalities came to the agricultural camp at Friday Bridge. Local farmers provided a variety of jobs at piece-work rates for which the students nominated themselves when and as they pleased. Their accommodation in an ex-POW camp and basic meals were provided at a small cost.

The gypsies were regarded as the hardest workers, often beginning early in the morning and slaving well into the evening; but one cannot generalise – there was Lotty, an oriental girl, who worked like a navvy during the whole of her stay, literally jogging down the rows to the check-out with her laden chips and missing lunch to pick as many as she could. But for the majority, the picking was a subsidiary income and they therefore took things a little easier. It was Dick's (a full-time worker) duty to allocate the rows, to check all the ripe berries were being gathered and that each row was completed before the picker moved on to the next vacant row. He would call you back if you had missed a patch and scold you if too much stalk was being included in the basket.

When 3.30 p.m. came round, the manager rolled up in the minubus to pay the students (they were paid by the day, as for many one day of fruit picking was enough) and drive them back to the camp. The scales were then removed, our totals collated and everyone started to drift home. It was with a feeling of physical tiredness and the satisfaction which follows, that we made our way home.

Often we would pass the 'cat lady' shuffling out of the gates on her two-mile trek home. She was a shadow of her former self; the animals she loved so much had ruined her – taking both her health and wealth. When we first made her acquaintance she lived in a house, with cats everywhere. Folk with unwanted litters would dump them on her doorstep by dead of night and as time went on, she could not feed them. Her diet consisted mainly of lettuce sandwiches and she would talk in all seriousness about the high nutritional value of the lettuce. We gave vegetables to her, but even the simplest she quizzed us on how to prepare and we knew they would end up as fare for the cats. Her body became so thin

and her skin transparent that the district nurse called once a week to administer iron injections. Her financial affairs became in such a state that she had to put the house on the market and sold the adjoining land to Eliah for his horse. She and her family of cats moved into a caravan in a corner. It was certainly a coming down in the world; her voice showed her to be the lady she was and rumour had it that she had moved in high circles in the better part of London from whence she came, but now she was obliged to earn what she could working on the land, and with her frailty it was not very much. Her movements were slow, she could drag her feet very few paces before she had to rest by the roadside, and so where possible we drove her home; it was not out of our way. When Clyde's snowstorm arrival featured in the local paper, I received a letter with a handmade gift for the baby 'for all our kindness towards her' - the present was a soft toy - a cat - stitched with childlike ability. It was a beautiful but sad gesture.

August gave way to September. The mornings would begin damp, the wet grass a reminder that it was time for the annual pair of new wellingtons. Spiders' webs hung with dew; the blackberry vines dripped water up our sleeves; raw hands picked sodden fruit and the berries weighed heavy at the first 'check in' of the day. The long school holidays ended and the women came later to work. The place seemed empty without their ceaseless chatter. The blackberry picking would continue a week or two more yet, before the women departed to the shelter of the carrot-topping sheds and the students to their studies in their respective countries. Rick and I were now moved onto other fruit beside the gypsies, doing the odd bits and pieces that it would be pointless putting a whole gang onto. We felt proud to be considered capable of working alongside these people.

First, there were the different varieties of plums to be picked. It often meant weekend work, because plums will not hang long. We gorged ourselves on them as we picked - and we soon knew when we had eaten too many! Rick would take the long ladder and I would use the steps - consisting of a ladder with just one leg, the purpose of which was for anchorage through a branch.

Once, there was a crop of cherries - we went in eagerly that first morning, but they were Morellos and gathered under-ripe too. However, after the initial disappointment, it was a pleasant enough occupation, plucking the loaded fruit in the same manner as one milks a cow and watching the little girls dancing about with bobbing cherry 'earrings'. This was 1976, but we were too far north for a good cherry crop - a sharp frost would always come at just the wrong time and so this orchard was eventually scrapped.

Rick and I felt hard done-by when we were given the headlands of Marjorie Seedling plums to pick. The trees, planted as a windbreak, were high and spindly, which made ladder work fearfully hazardous, added to which they leaned over a dyke. We could have counted the number of plums on our two pairs of hands and had to climb up twenty feet to get at them. We grumbled a week over those trees.

We enjoyed picking 'Cricket', the mature orchard behind the 'house'. It was enclosed by high hedges, a real secret garden, shut away from the hustle and bustle of the outside world. If the wind lay in the right direction, we might hear the church clock strike the hour or the sound of children at the village school out to play. The surroundings were of absolute peace and tranquility; birds would twitter all around us and rabbits hop about in broad daylight. It did have its disadvantages though, like sitting for an hour or more waiting for boxes. Also, because the trees were all so long established, big, old and gnarled, one could not expect a record week here. Around the edge Coxes Orange Pippins were planted - they did not do well on this soil, but the flavour was lovely.

Pears could be profitable, if you were put on the small Conference trees. There were King Williams and Du Comice too, planted every so often for cross-pollination purposes. I always thought these looked more the shape pears should look with their plump bottoms, rather than that of the Conference, so popular now. As a result of the dry summer, many pears were under-sized by EEC standards, and any under certain minimum dimensions had to be left on the tree. It was difficult to decide what to pick and what to leave. An additional bonus about this time, would be the clusters of mushrooms that sprang up each morning, if you managed to get to the right places first.

There was a variety of dessert apples. There were the Red Ellisons, a variety not in fashion now, rosy red like its neighbour Worcester Pearmain, but a soft fruit; the Russett, sister to the Cox but with a thick coarse skin, the Ida Red and the Crispin, a dual purpose apple that will hang until December. But it was Bramleys, over all things, where the money could be earned. The trees were

PW

an easy pick, the round apples would fill the whole palm of your hand, and in no time at all the boxes would be complete. We reckoned on filling five in a morning and four after lunch - but, of course, the price was reduced accordingly.

Soon it was October, the grass often wet, the wind cold and we would eat our sandwiches sitting in an empty crate. It gave us a feeling of being in tune with nature, living and eating our meals in the open, day by day, through the scorching days of summer into the cooler days of autumn; and now winter was just round the corner. As the season drew to a close it was not unusual to be laid off for odd days; sometimes because of the weather (there is nothing more miserable than picking a saturated, dripping tree); sometimes because of a backlog in the store; sometimes because there was nothing ready to pick.

Towards the end, there were the windfalls to be collected up. This is where the children came in useful; it is more tiring than you would imagine and certainly back breaking, stooping all day, avoiding being poked in the eye, losing your hat or getting your hair painfully snared by branches. Anything was collected, the decaying, rotting apples along with the wholesome, destined for the cider industry. One year, the farm had an urgent order for cider apples early on, and Rick and I were set to actually shaking the trees (which normally one was expected to treat with the utmost respect) to bring down the fruit with all possible speed.

Then the season was over for another year, and I was glad to be at home again, tidying up outside in readiness for the dormant spell, doing the housework that had been pushed aside for so long.

Jack was the foreman when we first went fruit picking. He was greatly respected by everyone and although his manner was one of meek composure, he could keep order. He would certainly have something to say if the fruit in your crate was bruised - after a few hours in cold storage fingerprints become clearly visible on apples pulled from the branches; a delicate lifting and twisting motion is required, transferring it as though made of meringue to the canvas bag over your shoulder.

Mr Selby, the owner, certainly looked the part as he walked round the estate in his plus-fours. His large old mansion is one of the nicest I have seen, with a walled vegetable garden, greenhouse and fruit cages. He would stop and talk to one and all. He wondered how long this sort of labour would be feasible, how long before machines would be perfected (like the strawberry machine already on trial) to do the task more cheaply, in far less time. Then this community occupation would be lost to the people forever.

CHAPTER 7

Paying the bills

When we set out for our new life in the country, we had no regard as to the financial situation – we thought we would be self-sufficient on our one acre of land. But, in reality, there were a multitude of bills to be met for general everyday expenses. After our rapid disillusionment over the condition of our cottage so soon after the move, it was obvious that money had to be earned, and quite quickly too. So we soon found fruit picking work for both of us in the next village. The income we reaped from the fruit picking, the strawberries, flowers, pigs and other bits and pieces amounted to a fair proportion of what we estimated to be our needs, but, after all, we did have a low standard of living. However, it was deemed necessary that Rick should find some form of work during the remainder of the year.

I conscientiously wrote down everything that we bought in order to ascertain where savings could be made. A look at an entry in our diary for 1975 shows our typical weekly expenditure:

	£		£
Milk	*1.54*	*Cheese*	*.69*
Cigarette papers	*.03*	*Onions*	*.10*
Sweets	*.12*	*Stamps*	*.16½*
Tea	*.11½*	*Tea*	*.63*
Toilet paper	*.36*	*Fishing tackle*	*.50*
Writing paper	*.23*	*Car-park charges*	*.10*
Yeast	*.09*	*Tobacco*	*.30*
Salt	*.44*	*Drinks*	*.80*
Sugar	*2.16*	*Refreshments at seaside*	*.57*
Butter	*1.25*		
			£10.19

Other expenses were the house rates at £36 per annum, the first electricity bill (covering cooking, a one-gallon water heater and mains radio) at £9 and petrol which was then £0.73 per gallon. On top of this was the cost of animal feed:

25kg dairy nuts	*£2.03*	*25kg wheat*	*£1.70*
25kg rolled oats	*£1.65*	*25kg layers' mash*	*£2.05*

One day in the autumn of 1976, we joined the staff of one of the carrot-topping establishments in the vicinity. Most of the women went carrot-topping when the blackberries were over and the other picking work became more physical, and the weather more likely to be unsettled. Herb, Audrey and young Audrey earned pocket-money in this way too, when their land was in order. The work continued as long as supplies lasted - from early September until February perhaps. It was just a large shed attached to a private house. Jade was able to play with the other children but could come to us when he wanted. Toilets and washbasins were available, but otherwise conditions were primitive; the workers ate their packed lunches on upturned packing crates and there was no form of heating whatsoever. The large doors at both ends of the building were always open wide to enable the fork-lift trucks to drive back and forth, bringing whole carrots and removing the prepared ones, ready for conveyance to the canning factory at Wisbech.

The work entailed drawing the whole carrots across a razor sharp blade embedded in the bench, in order to remove the green tip, out of which the leaf once grew, and also any damaged parts. Payment was by the completed sack of prepared roots, twenty-five pence at that time. The whole carrots were then fed mechanically down a chute. Proceedings were often held up for some reason or another and time meant money - the weather delaying lifting operations in the field was the main problem, mechanical break-downs were subsidiary causes. The amount earned could thus vary a great deal and it was also dependent on the size of the carrots. Numerous minute ones involved much fiddling, whereas a sack of large roots would be filled in no time. Carrots pulled in bad weather conditions would be muddy and unpleasant to handle, and another hazard of the job was the blade - working at speed, it was easy to catch a finger against it, as I very quickly learned. We only earned £1.50 that day, and that was for both of us!

Then, Rick was offered a lorry driving job with a general haulage contractor and garage proprietor. Though the hours were long and it would involve approximately two nights away each week, it was a chance to earn some money, and it meant

I did not need to return to the carrot-topping shed.

Rick worked for two and a half months, but it seemed much longer. Each week he would have at least one trip to Central London - in the belief that as he was from down that way (Royston) he would know the city! The other drivers were all local and were not prepared to do these runs. However, his trucking did take him all over the country, to Portsmouth, East Anglia, Yorkshire and once into Scotland. This was two weeks prior to our baby's expected arrival, and involved Rick being away three whole days and nights, so I anxiously hoped that the awaited day would not come too soon. A night in the cold, cramped cab was not a comfortable experience, so Rick would aim to reach familiar ground where possible - an aunt in London, another in Yorkshire, his parents' house near Royston - so that the lonely nights in the darkness of car-parks and lay-bys were reduced to a minimum.

The money we had been able to put by did not last long with Christmas upon us, and towards the end of January the kitty was empty. Rick rang the foreman of a firm of glass distributors, where he had regularly picked up loads whilst driving, and was consequently taken on to perform general warehouse duties. The work was tedious in the extreme; and, besides the pay packet at the end of the week, the only benefit was a pair of work boots and an anorak.

It was during his time here that Rick had his appendix removed. One Sunday he had a nagging stomach-ache all day and on waking next morning a sip of water made him vomit. I drove him to the hospital, where the offending part was quickly removed. While Rick became totally immersed in the routine of the sterile unit for seven days, I found myself with ridges to dig for the seed potatoes and generally caught up in the sudden rush of the spring chores. Audrey, always there when needed, looked after Jade during evening visiting times, but I considered little Dicken, only three months old and knowing only breast milk, too young to be baby-sat, and our presence on the ward was not very welcome. Rick was finally persuaded to leave the hospital, and after six weeks of convalescence he was declared fit to perform his lifting work again.

The activity of summer flew past, leaves changed colour and began to fall. Once again, Rick looked for a job. Looking under the agricultural column of the Job Centre's notice board, vacancies existed for stockmen on a duck farm, eight miles from where we lived. Rick decided to go for an interview immediately, with us and all our shopping also in the car. The very minimum of questioning was conducted through the bars of a stock building window; Rick hunched on the outside in teeming rain, the foreman and a group of workers sheltering from the rain on the inside. A dripping wet Rick climbed back into the car - he was to be a stockman as from tomorrow.

The work was repetitive, but Rick was always busy and there was no time to be bored. The hours were 7 a.m. to 4 p.m. followed by one hour's compulsory overtime. The eggs were incubated at another farm and transferred here as day old ducklings. When old enough to no longer require artificial warmth, they were removed to a pen in the open. At about eight weeks old, the majority were sent for slaughter; those chosen for breeding would later be moved back inside. On arrival each morning, the stockmen would collect the eggs to be taken to the incubation unit; only those of uniform size were suitable, the outsize ones, probably double yolks, were sold. Next, the fat poultry were loaded onto the lorry, destined for slaughter, handled by their necks with feet dangling. The slaughterhouse on the premises had been closed down as it did not meet EEC standards, so the farm had joined with two other duck producers to form a co-operative, with joint ownership of one satisfactory slaughterhouse. The remainder of the day was occupied in strawing down and mucking out.

Livestock have to be attended seven days a week - Christmas, Easter and bank holidays - but it was not really a bad job and he was working more or less out of doors. Then one Friday, he walked glumly in with his pay packet and said he had been laid off. Apparently he had mentioned in conversation he would be finishing at Whitsun, because of our strawberry crop. The boss had got to hear of this and declared that if that was the case he could go now. It left us in a predicament, but we scraped by as we usually do, and two winters later Rick was taken back on again.

CHAPTER 8

Our children

Although not concerning country matters, this chapter is a way of introducing our four children. It is they who make all we do and work for, purposeful and worthwhile. Jade was with us when we moved, and it was not until we had found our little piece of land, that we both felt the time was right for a second child. I conceived at about the date that the builders moved in. It was amongst the ensuing chaos, as we moved from gutted room to room, over rubble and builders' materials, living with grime and dust, that I had cravings for oranges and morning sickness lasting all day. The course of pregnancy ran beautifully smooth; the glorious months spent fruit picking in the village as usual. Ivy Cottage gradually took shape and I busied myself knitting bright baby clothes.

My expected date of delivery was 20 December 1976, so I was anticipating Christmas in hospital; but on the fifteenth, at about midnight, I began having mild pains and when 5 a.m. came, I whispered to Rick, snoring by my side, that I thought the baby was coming. He shot up, light on, and soon had Jade awake and in Audrey's care, next door. Then we drove through the frosty darkness, along the winding lanes into Wisbech. I think back on this confinement as a pleasant experience; relaxed and only at the hospital four and a half hours, before our second son, Dicken, was born, weighing in, as the smallest of our children, at 8 lb 6 oz. I was back in our own bed within forty-eight hours under the care of Nurse Marshall, with her old-fashioned but kindly ways, and there was much happiness in our little home that Christmas time.

Hoping one day for a family of four children, we waited until Dicken was fifteen months, before planning the third. It was in the spring of 1978 that my pregnancy was confirmed; the baby due in February. Much the same course of events followed - morning sickness, then another summer spent fruit picking.

It was on the fifteenth of February that violent winds blew up - bitterly cold with flurries of snow. We stocked up with animal feed, lest we became snow-bound - the force of the storm whipped across the waters of the River Nene, battering against the car parked on the quayside, as we transferred sacks into the boot from the mill steps. The storm rampaged into the night, and we awoke to blizzard conditions; a white world and grey skies. We tended the stock, breakfasted and wrapped Jade warmly for school; within minutes he was back - the bus was on the main road and could get no further, King John Bank was impassable - no school today! Rick busied himself with bringing in the day's supply of wood, while I washed and tidied up. Soon I realised that the baby would come that day.

Although it was not an emergency case, it would have been folly to have delayed in setting out for the hospital. As the car refused to start on such a cold morning, an ambulance was organised, but as it was now at Downham Market, fifteen miles away, it seemed wise to proceed any way we could, so we arranged a lift in a local farmer's landrover. With hurried hugs and kisses to the children, we began the bumpy ride. Two miles down the road, the ambulance met us, having almost become stuck at one point. Rick and I climbed aboard and it was 1 p.m. when we entered the hospital gates in Wisbech.

The contractions were mild and continued so throughout the day. When night came I took a sleeping tablet and slept for a while. The clock said four when I stirred - the baby would be born very soon. Because of the adverse weather outside, no midwife was on the premises. A student midwife, asleep in the lodge in the grounds, was summoned and Clyde arrived, a bouncing 8lb 15oz., a few minutes before 5 a.m.

We did not actually voice any decisions, but we were both content to stick at three children at this stage. However, I started feeling nauseous upon wakening and generally lacking in energy and vitality, and after visiting the doctor my fourth pregnancy was confirmed.

One spring Sunday, our little one already two days overdue, Rick borrowed our good neighbours' tractor and disc harrow, and prepared the half acre vegetable garden. I busily sowed lettuces and radish seed, cauliflower and winter brassica. I was pleased to have these safely in the ground, not knowing when the baby's arrival would halt further progress on the land. That evening I helped Rick to bottle his home brew and then we retired to bed.

I awoke at about 2 a.m. and after feeling mild discomfort for half an hour, concluded this was to be it and woke Rick. Being the fourth time round, we thought it best not to delay. On arrival at the hospital, I was hurried down corridors to the delivery room. Within a few minutes, the second stage of labour was under way, and at last our baby came into the world, and, such joy - a girl! We shortlisted three names - Ella, Megan and Bryony - and finally chose Bryony. She, like the boys, was of a good birth weight, 8 lb 10 oz.

Now our family was complete - Jade, Dicken, Clyde and Bryony. How fortunate we are to have such a fine, large family.

CHAPTER 9
Mr Pogle

One weekend soon after we had settled in at Ivy Cottage, Mick and Theresa came to stay. They had been close friends from our Royston days and in fact Rick and Mick had lived in the same street since early childhood and grown up together. It was decided on Sunday that the men should build a chicken coop. There were numerous tea breaks and a lengthy disappearance of the two carpenters to the Wheatsheaf for a cool pint, but surprisingly by the end of Sunday, we were summoned to admire their work. A hen-house stood in the area Rick had already permanently fenced. It was small, but very skilfully put together out of the tongued and grooved boarding salvaged from the bedroom floor at Field Crescent. The door slid open for ease of cleaning and access to the nesting boxes, and the hens came in and out by means of a ridged path, through a hole at the front. There were perches for the birds to roost on at night.

The accommodation ready, we could now purchase some hens. We decided to go over to Swaffham the following Saturday, where a large auction market was held, selling everything from vegetables, trees, rabbits, poultry, goats, timber, vehicles to furniture and junk. The prospect of bidding somewhat alarmed us so Herb volunteered to accompany us, happy to do it for a drink outside the pub, which did a roaring trade all day, in the market yard. He liked nothing better than a day out at Swaffham and his son took him along when he went once a fortnight. We soon became the owners of four Light Sussex hens and two Rhode Island hens for the total of £4.08, plump for the price. There was more competition for the fine Rhode Island cockerel on which we had set our hearts. Eventually, the bird became ours for the quite high price then of £2.05. We called him Barnaby.

Once home, we stood in fascination for ages, just watching them contentedly scratching and pecking. We were always popping into the run to 'check for eggs' in the days that followed. When we had company, we would produce new laid eggs for tea – so proud were we of the deep yellow yolks of poultry fed on unlimited quantities of greenstuff.

When Rick's parents came to visit, they presented us with six extraordinary-looking birds, Sterling Rangers, on point-of-lay, and very shortly we were self-sufficient in eggs. In fact, these are the only hybrids that have done well under our system.

The chicken run soon started to look like a chicken run – bare compacted soil with tall patches of nettles. The strawberry block situated next to it benefited greatly from the leaching, thriving in the acid conditions.

Winter arrived; not a bad one by any means, but the old hens started dying and as we were vegetarians, we buried them under the newly established fruit trees along the bank, to fertilise them. We were quite sad about it and gave them a sort of funeral, placing a posy of flowers on their graves. The rats, or perhaps Nooks, Audrey's dog, promptly came and dug them up again. Nowadays, we would have quickly necked them for the pot at the first sign of frailty. Quite soon we saw that to be profitable egg-producers on whatever scale, we could not carry ageing hens, sick birds or any others that did not produce their quota of five eggs a week (allowing, of course, for a rest occasionally) and we considered it our responsibility to deal with the culls appropriately, consuming the resulting protein ourselves.

Barnaby had proved to be a virile rooster, but as he got older and his spurs longer, he became vicious in the protection of his harem. Jade would no longer check for eggs and it was certainly with some trepidation that I crept in – once your back was turned, he would fly at you. So he eventually became our Sunday dinner – even the toughest bird is appetising after prolonged casseroling in a slow oven. Mr Pogle was bought, also from Swaffham, as our replacement cockerel. He in no way measured up to Barnaby's arrogance and grandeur, but he was still an attractive bird with multicoloured feathers.

Soon afterwards, we were given a silkie bantam, a miniature cockerel who Mr Pogle would have eaten alive. We did not particularly want banty crossed hens, but he was a lovely little thing, so we called him Jade's pet and let him range free over the remainder of the land for his own safety. One morning Jade rushed in hysterically – the door to the hen-house had blown open in the wind and Mr Pogle was murdering Jade's cockerel. I rushed out

to separate them. The little silkie, his downy feathers splattered red, slunk down to his perch on the rafters of the goat shed, where he spent several days before he deemed it safe enough to venture out again.

Once, we bought ten old ex-battery hens for twenty pence each, to put in the freezer. They laid eggs in the crate before we had time to kill, pluck and dress them, which tempted us to offer them a reprieve. But we feared they would not survive in our wild elements, only knowing warmth, confinement and electric lighting, and so we did the honours without more ado. We never found such a bargain again and I have always regretted not giving them a chance.

We had read that bantams make excellent broodies and so we went to purchase a dozen from a breeder. There was a large variety of types, but we had no choice in the matter. Catching the poultry, housed in large sheds, was so difficult that we just had to be satisfied with those she did manage to get hold of. The hens appeared to be old birds to our inexperienced eyes, they laid an egg once in a while, and were the culprits of the introduction of scaly leg to our flock.

However, when the season came round, they did prove themselves at least adequate sitters. Shut up, as they were, all together in a run, if one got off to stretch her legs, have a scratch about and a sip of water, when she came back to her eggs, it was to find another had taken her place. Five broodies reared five chicks between them; some eggs were deserted, but many others were infertile.

We put two more cockerels in to help tread the hens, although one cockerel should have managed to cover at least twelve hens. One flew straight out again and evaded all attempts at capture, although he was often sighted around Greg's yard.

We purchased an incubator because we thought this would be a more reliable means of obtaining replacement stock. We bought in seventy eggs from pure Rhode Island stock – the fertility rate was poor, thirty-three per cent, and it was a mere two dozen chicks that hatched. Conveniently, we had two broody hens at the time, so gradually we introduced them, giving twelve to each. Soon the more dominant of the two ousted the other and collared the lot. It was amusing to see her attempts to shelter her very large family under her wings, particularly as they grew.

The artificial apparatus had been much more reliable than the bantams in incubating the eggs and we use this method annually now to renew our flock - culling out unprofitable hens in the autumn. We try to use broodies as mothers though, as they do a far better job. The infra-red lamps involve a lot of fiddling about - renewing the sawdust bedding frequently and paying close attention to the hygiene of the hoppers, to avoid coccidiosis (easily spotted by the hunched up appearance and soiled hind-quarters) which spreads rapidly.

The pen must be safe from predators, with special thought to the netting size - we have had cats remove chicks through what we considered small mesh. Also, the size of the pen is important - a pen that seemed small enough to keep the heat in and the chicks warm (chicks huddled together indicates too cold conditions; chicks scattered far apart indicates overwarm conditions) can soon become too cramped and should be extended as they grow. We once realised this too late - the chicks were attacking each other. A spot of blood (a red strawberry has the same effect) will send chickens crazy and they will peck and peck at it until it is gone or removed from them. Chicks can also drown in the smallest amount of drinking water, so proper hoppers must be provided. They must be hardened off gradually to the outside elements when the heat lamp is switched out, in the same way as bedding plants raised in the greenhouse are before planting out.

Since we moved to Stickford we have, as a rule, allowed our hens to free range. They have been limited in no way, but usually stay on our seven acres. Occasionally, they have gone up onto the road and been run over, and for a while a few would wander through the hedge into our neighbour's field - this is the only impediment in the security system - the only thirty yards of the circumference of our land not surrounded by the restriction of dykes.

We have never found our hens to lay well in the shorter days of the year. There is usually a flood of eggs in the spring, when even the oldest hens will produce, tailing into a moult during July and August; then a fair amount of eggs in September and October, although not as many as in the spring; few eggs from November through to the end of January, then the cycle starts all over again. We aim for an early broody February and March when the pullets should be laying, but in order to have an adequate supply of eggs when the long evenings are with us, we preserve two big crocks of eggs in water-glass.

One year the hens were in full lay when we confined them in a large barn in order to keep them away from the strawberry field. The eggs promptly dropped to three a day. Tell-tale signs of broken shells, whole ones sticky with yolk, indicated that the hens were eating their eggs. We tried everything – putting sacking over the nesting boxes, filling an empty shell with mustard, painting the shells with the varnish used to prevent children nail-biting and supplying endless buckets of weeds as an anti-boredom measure and to provide nice yellow yolks for the non-existent eggs! But it was not until the hens ran free again that the problem ceased.

We have decided that in future, if we are to make a real go of our place as a market garden concern, then the hens will have to be penned - they scratch all the sown seeds out of their rows, peck chunks out of the root crops, unearth the potatoes, and do a lot of other damage. The ideal situation is a pen which can be moved around, to fold the poultry over the pasture.

We have pondered the question of how 'free' is free range. Our hens are in no way restricted; acquaintances also have poultry, which they keep in an outside run in their small garden, fed on proprietary pellets, which they too regard as free range. To comply with the Trades Descriptions Act (as defined in Shrewsbury Crown Court) a free-range egg must have come 'from a hen with regular daytime access to land from which it obtains a significant proportion of its food from the natural sources provided by that land'.

CHAPTER 10

Dragging their stakes behind them

One Friday, relaxing with the local paper as we munched through my favourite lunchtime snack of cheese and raw onion on toast, an advertisement caught my interest: *kid goat for sale, £15.* Milking stock had not yet been discussed, but to be self-sufficient it was perfectly obvious that we must have a goat - a cow was out of the question in our set-up. As usual we rushed straight in; there was no telephone number given, so we just got into the car there and then and drove to the address given some fifteen miles distant. The street was a row of terraces. As we made our way through the archway to the back of the house, we knew we were in time. A young goat, Nubian type (crossed with a Saanen we were to discover later), was tethered on the lawn.

The sight of the garden, and of the neighbouring ones on either side, should have warned us of things to come. The area was one of total devastation - nothing edible remained; the enclosing privet hedges were totally devoid of any green vegetation, and the lawns were nibbled down to the earth, bare brown patches of soil mingling with the dry crisp yellow, where the grass had failed to rejuvenate.

It was terribly exciting - our first real animal (we had only acquired the hens to date). She slotted easily into the back of the estate, as did the adult goats who came later. Once home, we staked our goat out on the lawn. Tethering is not advisable with kids, but Ruth (as she had been named) was already well broken in. We piled seeding lettuces, outer cabbage leaves and peelings in front of her and generally fussed over her. She soon demonstrated the fallacies of the supposed eating habits of the goat. Contrary to popular belief, they are fastidious animals, picking over the contents of the hayrack, ignoring the bedding straw or soiled fodder on the stanchion floor. Unless half starved, they will not act as lawnmowers either and tend to walk round and round in circles, complaining loudly and tangling their tethers, when just offered boring garden lawns to 'graze'.

We discovered that, in fact, goats are browsers rather than grazers, so as she matured and the herd increased, the goats were tethered on the wide sides of the roads and dykes that ran right through the village - here they would nibble with relish the herbage, especially the cow parsley, and happily munch the long grass. While in the public eye, they were forever being fed titbits, and cars would draw up for children to pet the animals. I should add, however, that tethering is not infallible; many is the time that, late in fetching them home, we have heard the jingle of bells (each milker in the herd was fitted with a Swiss goatbell) and looked out to see the goats sauntering down the lane, dragging their stakes behind them!

In the evenings, Rick would go out with his scythe to mow grass to fill their hay racks. We squeezed in a late crop of turnips that first summer for Ruth and we specialised in fodder beet in the years that followed. We found we could obtain an unlimited, free supply of sacks of carrot tops from the local carrot-topping factory, which provided a particularly succulent change from their dry diets in winter.

Ruth possessed a fine pair of horns and when a goat has horns, it will use them, so we felt we would be happier with them removed. Disbudding, as it is known, is usually done during the first four days after birth with a hot iron; by law under anaesthetic. We have calves dehorned by this method too, but their growth rate is slower than that of the goats and it is best done at four to six weeks old. Before Ruth could be tackled in this way, her long growth of horn had to be sawn off. Until the site healed, vigilance had to be kept for blowfly attack. Some goats are naturally hornless and these should be crossed with a horned male, otherwise there is a strong risk of hermaphrodite kids resulting.

We wanted to do everything correctly for our goat, so we took out every book written on the subject and bought all the leaflets published by the British Goat Society including the dairying side, although it would be some time before our knowledge was called upon. During our early days at Walpole, the Fenland Goat-keepers' Society was set up. In our quest for knowledge, we were to learn much from the experienced goatkeepers and smallholders there; and the society went from strength to strength.

As the days went by, we began to realise that rearing a kid was going to be an expensive means of acquiring a milker (which were fetching £40–50 at that time), on top of which Ruth could not be mated until her second autumn. If starting out again (as we put into practice when purchasing our first cow), we would have bought a down-kidder/milker.

We were impatient now to have our own milk supply. Milk had risen to the grand total of eight and a half pence a pint by that autumn. We dreamt of hard and soft cheeses, butter and goat's milk yoghurt, for which Rick had a soft spot. So we acquired Betsy, on trial only – on the condition that should we choose to keep her, the kids she was carrying would belong to the previous owner. Betsy was pure Anglo-Nubian and closely resembled a camel. She gave us two precious pints a day. I never did come to enjoy goat's milk as a drink on its own, but it was usually undetectable in tea and cooking, unless turnips, kale or other brassicas had been fed too close to milking time, when they would taint the milk. Cobalt deficiency also gives a bitter taste to the milk – easily corrected with a dose of cobalt salts. However, Betsy did rather more than just reduce our milk bill, and after five weeks she had to go. I have never met a nastier goat. A certain amount of scuffling is understandable, it happens amongst all groups of animals until a pecking order has been established, but Betsy was plain vicious.

In February, Pat, a friend of Rick's parents, gave us a goat. 'Nan' quite defied description – she was a real mixture of types – though to say she was a British Alpine type would be kinder. She was giving very little milk when we took her on, but within days her perpetual wagging and bleating signified that she was on heat. We took her to be served and in July, with the best grass gone, two kids were born, a male and a female. Nan proved to be poor in every respect, and before long we advertised her, bartering milker and female kid for thirty bales of hay.

The billy went on to be the first of many we fattened for the freezer. With plenty of grass, it can be a worthwhile exercise. The meat we found to have a distinctive quality of its own – a stringier texture than lamb, with a more 'gamey' flavour. As tasty when roasted as when curried, the carcass lacks fat, so care must be taken whilst cooking to avoid drying the meat out. The skin is cured by immersion in five per cent formalin solution for five days, then it is stretched out and nailed on a board with plenty of air circulating

around. After it is completely dried, the skin side is rubbed down with sandpaper, then with lankroline. The finished product makes an excellent Christmas present.

In the spring we bought Trixie, a British Alpine, already milking, and made the gesture of cancelling our delivery of milk. We produced butter that looked like lard, obtained by churning whole milk in a glass Blow churn (the cream is harder to separate from goat's milk than cow's) and soft cheese by leaving the milk to sour naturally, which it eventually did or turned mouldy! We still lacked the quantity required for hard cheese.

A good goat is not one who gives record quantities at the peak of her lactation (said to be six weeks after kidding, her peak lactation generally the third), but one who remains constant throughout the winter. The yield from our goats always seemed to tail right off by about Christmas. Because of this dilemma, when we bought the freezer, we froze the surplus spring milk; stirred after thawing it would reconstitute. The cow is a very different matter - we have a hard job drying her off at six weeks before calving, still giving a gallon per day.

Autumn came and with it the mating season. Ruth and Trixie came on heat within days of each other, which is very common - the smell of the billy on one goat returning from service will often bring her companions round. Similarly if a goat fails to show signs of coming into season, a waft of a 'billy rag' often does the trick. Trixie and Ruth were taken to service, and so it was that in April each produced a kid, the former's a billy, the latter's a beautiful Nubian lady we called Bracken. She looked a picture sat gracefully beside her dam in the spring sunshine, and we would spend hours watching the two kids frolicking, skipping and leaping as kids will. Billy, as our males have always been called, we knew was destined for the freezer, but Bracken was special. She qualified for entry in the supplementary register of the Goat Society and we intended to rear her, so we allowed ourselves to become fond of her.

As the days went by and the kids filled out on their four pint daily milk ration, it became obvious that something was amiss. Bracken had developed a hobbling gait. On seizing her to investigate, we discovered her knee joint would not bend completely. We started to run up a large veterinary bill - the more she grew, the more severe the problem became, until one Saturday morning we took her to the vet at Long Sutton and did not bring her home with us. It was a subdued journey; tears rolled down my cheeks and Jade sobbed loudly.

The following year we had a much more prolific crop of kids from Trixie and Ruth. Ruth gave birth to triplets one night - a surprise for us coming down in the morning, especially as they had arrived a week early. There were two females, Carob (a chocolate colour) and Bracken II, and a male. Carob was so tiny we feared she would not survive. Ruth was not a good mother, so after a day of bleating and kicking the offspring were removed for their own well-being and bottle-fed (which we do at four days anyway). Much patience and perseverance was necessary with Carob, but after many feeds, with her cradled on my knee, stroking her throat to bring about an involuntary swallowing action, she learned the art of sucking.

Since early morning, I had known Trixie would kid that day - a stream of mucus dangled behind her as she restlessly moved about. Looking the matter up, the book claimed this signified an awkward kidding. By late afternoon, labour was well under way. She panted and pushed, walked around, puffed again and again - but nothing happened. Eventually she came and laid down her head in defeat upon Rick's lap. She needed assistance. Rick went to clean up in preparation for an internal examination to ensure that the kid was presented correctly. I went for Eric (a local goatkeeper), as this was our initial exploit into the midwifery field. By the time I had returned, Rick and Audrey were smiling smugly in the yard. Rick had been successful in extracting the female kid (that was to be called Bryony) but there was still another there. Rick and Eric struggled for two hours trying to bring this kid into the world. We thought there would be no hope of its being alive. Rick telephoned the vet, who came along as soon as he was able and extracted a male kid - and it was still breathing.

Trixie was not a heavy milker. We did not feel her yield merited her keep, so that summer we sold her and a female kid, and bought in another milker from Eric. Kate running with the herd was only giving two pints a day in August, but with a little individual attention this rapidly doubled. Kate was the Saanen I had always wanted and she had an impressive pedigree, seventy-five per cent Mostyn. She had a beautifully placid temperament; she would just plod along. I always thought she should have been born a cow.

By 1979 our goat numbers reached their biggest yet, boosted by both goats giving birth to triplets – Ruth two females, Rosemary and Rue, and a male; Kate two males and a female, Dill. We had registered our prefix with the British Goat Society as Disilo, which is a Romanian gypsy word meaning the dawn; hence her name Disilo Dill. We killed two of the billies off young to help the desperate milk situation. Ruth gave eight to ten pints daily and although Kate was producing a spectacular yield of fourteen pints at peak, six kids at four pints each day meant a severe shortage for the house!

Not wanting our dreams of hard cheese-making to be shelved for yet another year, we went out and bought another milker and kid (Toggenburgs) Gwen and Florence, intending to sell Florence on rapidly. Our goatshed walls bulged – there were now twelve goats: three milkers, Kate, Ruth and Gwen; two goatlings, Carob and Bracken; and seven kids. But soon we began steadily reducing the herd as we thought about moving and in the end only Bracken, Kate and Ruth moved to Stickford with us.

Kate and Ruth were both dry and obviously well in kid, and each three week cycle I claimed Bracken had returned, but Rick remained unconvinced and so she was not re-served. In the April of that year, Ruth gave birth to a pair of twins while tethered along the roadside. We had checked her five minutes previously and she had been browsing in earnest. The female kid was bartered for a cow halter; the male went into the pot young. Kate again produced triplets after a prolonged labour; a stillborn male and two females, Sorrel and Sage. Disilo Sage was sold at four days old.

For a couple of days prior to kidding, Kate appeared to be unsteady on her feet. Not dramatically so, only once or twice she stumbled onto a knee as I walked her up Scarborough Bank. It became increasingly difficult to get her out at all in the mornings. I began only to tie her to the hedge on our side of the road, nearer and nearer to her stall. We sought veterinary advice, and chronic constipation was diagnosed. She was given enemas, multi-vitamin injections and steroids, all to no avail. We tried to tempt her with all the herbs and titbits we could think of, and dosed her with liquid paraffin. She grew thinner. It came to the stage that we would leave the goatshed door open for her to venture out if she chose and she never did. In fact, she seldom rose at all. Alan and June Smith were experienced goatkeepers, and knew of a greatly respected vet in King's Lynn who specialised in goat health, so we let them take Kate home with them. We had done all we could and we wanted the best for Kate. It was not much later that June rang to say that poor Kate had died. Had we known her days were so numbered, we might have let her die peacefully in familiar surroundings. It is sad that Kate's full potential was not realised until her end, our Kate (Bonwycks Manetia) the fourteen pint a day milker.

But Sorrel was still with us and we understood that she was something out of the ordinary, so we entered her into the goat section at the Revesby Fair. This was my initiation into the strange world of showing. A friend lent me his white coat, so at least I looked the part. Everyone seemed to know everyone else. Cars were filled with branches and browsings for the pampered stock, who had all been shampooed for the occasion. Sorrel looked positively grey. It became obvious that the condition of the animal was not all that counted - the animal had to be prepared to order too; to walk correctly when led and stand correctly when handled by the judge. Sorrel and I did nothing right and I was severely reprimanded for it, but the judge liked Sorrel; at four months there was already milk in her udder. One glaring fault he explained to be the biscuit shade in her coat, which we should disguise with chalk at future shows (not far short of cheating to my eyes), but Sorrel was awarded a red rosette, as first Saanen kid in her class.

Sorrel's success went to Rick's head and he anxiously found out when the next show was so that she could be shown again. At the Fenland Goat Show in September, the last one of the season, Sorrel was the only competitor to turn up in the Saanen kid class, so the show secretary amalgamated two kid classes, Saanen with British Saanen. Unfortunately this put Sorrel at a disadvantage; she appeared small against the others, British Saanens being a naturally larger breed of goat. But it was a large class and Sorrel received a rosette for fifth place.

At Stickford, we had established a permanent pasture and the goat's milk was now in addition to Blossom's (our cow) three gallons. Ruth saw us through Blossom's first dry period, the two months between her first and second calving, but on comparing the two we came to admit that we preferred the cow's milk in every way - taste, quantity, the cream and its diversity of use, therefore, we bought Cello, to cover Blossom's second dry spell, but basically to be a suckler. So, with some sadness and regret, we sold Ruth, but we made sure she went to a good home.

During the strawberry season of 1981, we advertised Sorrel in the local Goat Society newsletter and she was sold to the owners of an antique shop as their first goat. We went back later to discover how they were getting on. Sorrel was in no way deprived of attention - her every whim was her mistress's command.

Our own goatkeeping activities ended with Sorrel, but each year we easily come by a billy goat or two to raise for the freezer. With the huge influx of goat owners, female kids are now not easily disposed of, and so the chance of these too comes our way. At the moment, we have a billy that should not have been. His mother came on heat out of the breeding season, but May or not, the billy did not mind! The kid arrived on 2 October and is called 'Jaiz Amazing'.

CHAPTER 11

No need for a lawnmower

We had been in the country for about a year. We had acquired chickens for eggs, goats for milk and, now that we were no longer vegetarians, we decided to breed rabbits for meat. Rick hurriedly put together some hutches, in whatever materials happened to be at hand. The next day we bought a New Zealand White doe plus litter, together with two other does and a buck of the same breed from Swaffham market. We thought that three does and a buck should keep us well-supplied with meat throughout the year, but they were not prolific. The saying 'breed like rabbits' became a joke in our house.

Next, we introduced a young Flemish Giant buck, a magnificent pure-bred creature, bought from a local man who produced them for meat as a hobby. The idea was that the New Zealand Whites would throw the large litters and the Flemish Giant would increase the body weight. Still no offspring followed, and it was not until the male was six months old that any of the does showed signs of being in kid (emitting a high pitched squeal when introduced to the buck, and refusal to mate). We then realised that the larger breeds must be slower to mature sexually. Now the breeding was well away.

Rick constructed a block of six fattening hutches, which a friend christened 'Death Row'. Maybe the inhabitants understood, for we were continually running here, there and everywhere, catching escapees. As our rabbit keeping progressed, some of the dark does, the result of crossing the two breeds, were put back to a New Zealand White buck. The resulting offspring were either white, or dark like the doe, and some were born resembling the wild rabbit. None of the young produced had mottled markings as one would suppose.

We found rabbits ideal on the small acreage at Ivy Cottage. They were fed mostly on weeds gathered from the roadsides, grass scythed with the goats' daily ration, and any greenstuff we could scrounge. Living within the community, the latter was enormous.

Audrey collected all her household scraps; the villagers gave us their brassica stalks and grass cuttings and the allotment holders their waste products. It was common to be given an allotment of lettuces, cauliflowers and brussels stems to clear for the animals and we always retrieved what we could use for the house first - generally this was a great deal. In addition to this, we removed the backs from two rotten wardrobes of Maud's, replaced them with wire-netting and ran the rabbits over the lawns; thus these were fertilised and we had no need for lawnmowers. We could feed the rabbits very cheaply in this way, and only in very cold weather did we introduce any grain (rolled oats) to their diet. Once we lost a hutch of rabbits through bloat after feeding them wet, rich greenstuffs, but generally there were no problems.

Rabbits are dirty animals, but instead of carting their waste to the tip, we used all their muck and by feeding the weeds and produce of other gardens, we were buying in fertility. A third by-product was the skins. These we cured by the formalin method used with our goat skins.

After we moved to Stickford, rabbits became less important. I still like to keep a pair - rabbit, roasted then chopped into a white sauce, maybe with a pastry lid, is a very tasty meal. But we can now raise other meat just as cheaply, with four acres down to permanent pasture, and with less demand on our time; for example, Jersey bull calves and lambs which spend one summer finding their own food, after an initial start on milk. The cry 'catch that rabbit' has become 'lambs are out again!' And the muck from a cow wintered indoors makes the former 'mountains' from our rabbits seem insignificant, but rabbits were good for us at Ivy Cottage.

CHAPTER 12

'She's just eaten £20'

We had read what an efficient land reclaimer the pig was, fertilising the soil as she rooted out tree stumps and perennial roots; how she would devour all the household scraps and grow fat on them; and the vital part she would play as a link in the dairying cycle, consuming vast quantities of skimmed milk. This is our version of how in practice we came to reject all these theories, at least partially.

We now already raised small stock for meat alone - cockerels, goslings and rabbits. After eighteen months of husbandry, Grace and her brother arrived. We brought them home in the back of the car from an energetic lady smallholder over in High Norfolk. They were only weaners, eight weeks old, but we made a song and dance over moving those two small frightened creatures, each no heavier than a bag of meal, from our vehicle to the little log cabin, a hundred yards over the lawn!

The sty, with access to a small netting-fenced run, had been constructed of offcuts, with sacking hung over the doorway to cut out draughts and a bricked, but not cemented, floor. This hut looked really attractive - but was soon lacking its canvas door, and the bricks were rooted-up in favour of a deep excavation where the two pigs would lie. Very surprisingly the run kept the pigs confined well. We had heard extraordinary tales of the escapological traits of the pig, but, in our experience, she is not nearly as bad as the goats and heifers!

We kept the hog until we thought he looked the right size, which happened to fall in August. Then he was carted off to a butcher in Wisbech to be slaughtered and split. Our 'porker' weighed 180 lb dead weight! We bought a fourteen cubic feet freezer to cope with all the pork and put some down in a barrel of brine for bacon. Luckily, the blowflies avoided us. As each joint came out of the pickle on its appointed day, it was hung up in muslin bags from huge meat hooks from the kitchen ceiling. In a cool dark corner, it will keep indefinitely. Instances of hams being uncovered a hundred years on and still deliciously edible are not unusual. Herb in his younger days would hang his in the attic for three years before touching them. Burying it in chests of sawdust is another old method of storing the meat.

Brine Recipe

For 100 lb meat:	2 oz saltpetre
8 lb salt	5 gallons water boiled, then cooled
2 lb sugar (dark brown)	

Method Leave in brine barrel, weighted down by a wooden board and bricks, for 2 days per pound small joints, 4 days per pound large joints.

Dry salting is another method of bacon curing, but it has always seemed more trouble to us. A pig can be rendered inedible by oversalting in this way. We just eat our hams salted; one day we may get round to building a smoke house, particularly as we have acquired a taste for smoked fish too - or perhaps we shall move somewhere with a suitable chimney to hang the joints in. Bacon is about the nearest we get to convenience food. Thick rashers, sliced off with the carving knife - soaked a few minutes in boiling water to remove the excess salt, they are ready for cooking. A large joint boiled, eaten hot the first day and served cold for several days after, is easy summer cooking for busy periods.

The children are very fond of sausages, so as well as utilising the odd bits and pieces, we mince up the half shoulder too. Rick's Great Uncle Cliff was a butcher at Sutton Bridge, a couple of miles from us on the Marsh. He gave us this special sausage recipe.

Sausage Recipe

20 lb pork	½ oz pepper
3 lb rusk	3 pints water
½ oz sage	5 oz salt

Method Mince meat then combine with other ingredients. Attach funnel and pass mixture through the mincer and so into the skins.

We always use the small intestine from the previous pig killed off six months before, and reverse, clean and scrape those from the present pig, to put down in salt in readiness for the following one.

Brawn is made from the head and trotters, which are half covered with water and boiled until the meat falls from the bone.

The inedible bits (bones and teeth) are then extracted, and herbs and seasoning added in plenty. The semi-liquid is reheated and finally it is poured into containers to set, where a layer of fat rises to the surface and will preserve it. We pop all the remaining fat into a tray in the slow oven to render down. The crispy pieces left afterwards are called pork scratchings and Rick and the children fight over them before the cats get a look in.

Rick sometimes makes pork pies, using skills recalled from his catering days, with the apple corer hole left in the pie top, down which to pour the liquid after baking to form that special 'jelly' on cooling. We are indebted to Audrey for this hint.

Pork just becomes everyday fare to the smallholder, but what a wealth of variety the pig provides - haslet, black puddings, chittlings (the stomach and the membrane holding the small intestines together - chopped, boiled and then fried); the special suppertime treat of tender loin (the equivalent of fillet steak in beef) roasted foiled in the coil - the beautiful viscous membrane resembling a lady's lace shawl; and the offal.

The first time Rick jointed a carcass he followed a diagram in a book, but with practice his speed increased, as did the rate at which we dealt with the meat. Initially it seemed to be hanging around for days in various forms - this pan for mincing into sausages; that basin of skins still to be cleaned - but it is now rapidly disposed of.

At Stickford, we are fortunate in having Gordon, a licensed slaughterman, who comes round to shoot the animal with his humane killer on our own premises. In this way, it knows nothing. A copper of boiling water has to be ready for his arrival, and he is an early riser. If the water is at the right temperature, the hairs will scrape off easily when the body is immersed in the tin bath. We now only kill when there is an 'R' in the month, excluding September and the early part of October, and so we are safely able to hang the meat outside from the beam in Rick's Dutch barn for a day to set.

Well, that is the meat side of the pig enterprise. There are laws against the sale of meat, unless it has been slaughtered on licensed premises, so if we do run a surplus animal on for meat, we sell direct to our local butcher.

*   *   *

FW

We make our pigs pay by the sale of weaners, which takes us back to 1977 and Grace. Her brother departed, the next step was to get her in pig. Greg, a few doors away, kept a boar to cover the sows he housed. When on heat, Grace could smell Fred. It was all so simple; open the sty door and Grace would hurriedly trot out, up the lane and round the corner, her pace increasing the nearer she got, until she would break into a gallop, dragging her attendant along by her rope. When Fred went to the processed meat factory, never again did we have a sow come on heat so strongly and from then on, Grace had to be transported to a boar six miles away.

A problem now arose in that to move a pig other than to slaughter, a movement licence is required and, as a heat period is not always foreseeable, the necessary papers cannot always be obtained in advance. It is no fun trying to hold an amorous sow in while the paperwork is attended to, particularly when the Diseases of Animals Inspector who despatches the permits is at King's Lynn market, ten miles away. To be fair, he was always most helpful. Once, Grace smashed her sty door in her passion and I had to spend two hours tapping her on the nose. When Rick finally received my SOS, he had to collect the licence before I could finally be relieved of my charge. On this occasion, the Inspector met Rick in a car-park half-way.

With Grace now in pig, we turned our thoughts to land clearance. There had been two blocks of ancient unproductive gooseberries, underlying with couch grass when we took over. The smaller block we cleared by determined effort and hard work - a whole winter spent digging. Now with a pig to do the work for us, we should be redundant in reclaiming the second block; this is where the first theory was quashed. We divided the area into three pens, each 30 feet by 30 feet enclosed with pig-netting - blocks of a practical size in order for her to be concentrated enough to clear efficiently. A temporary outdoor sty was put up in the first pen (we did not intend her to farrow here in the open). When her task was completed, Grace and sty were moved to the second pen. Then the first pen was seeded. Geese, goats and a lamb always kept on top of this grass, so maybe we would not have noticed any couch still present, but it seemed a good enough job had been done.

This was not so with the second pen, which we wished to cultivate. Grace enjoyed the couch roots and did a good job in clearing them, but at the same time she would dig huge dips, bringing all the sub-soil (sand) to the surface - ruining our beautiful silt land. From our readings on the principles of organic husbandry, we had come to realise that all the life force, all the beneficial bugs were in those top few inches (hence the no-digging school) and Grace was destroying this structure. She did not even completely root out the gnarled bushes, only loosened them somewhat. Another theory was also disproved - she did not fertilise as she went. Contrary to popular belief, pigs are very clean animals. She set aside a corner for her toilet, and so the resulting dung heap still had to be spread in the same manner as from an inside sty.

When Grace moved on to the block of artichokes already established in the third pen, we had a terrible task cultivating the vacated pen. It had first to be levelled out. The soil was rock hard - the Merrytiller just skimmed the surface. Ironically, it really needed hand digging after her, but there was no time if it was to be cropped that year. There was no access for a tractor and plough onto the area, so we did the best we could; then planted it up. Obviously the roots of the little seedlings found the earth as

impenetrable as we did; nothing did well there that year, even the marrows and pumpkins in the corner where the dung heap had been. Few weeds grew though, so perhaps she did some good.

For health reasons, the law stipulates that pigs should not be kept within fifty yards of a dwelling. There should have been a fourth pen of couched gooseberry roots, but by this ruling, Audrey's house came within the limits and so we had to hand dig it.

As Grace's time drew near (a pig's pregnancy lasts three months, three weeks and three days) Rick began constructing a farrowing house out of breeze blocks. It was his first attempt at bricklaying and looked like it too! Measuring 12 feet by 12 feet with concrete floor and felted roof, it cost the phenomenal price of £100 then. We were dubious about our first farrowing; warnings rang in our ears of sows eating their first born, thinking it was the cause of all the pain, and then lying on and smothering several more. Rick fitted farrowing rails and a creep area with an infra-red lamp, where the piglets could go to nibble meal unmolested by Mum.

We did not crate Grace, but removed each piglet and placed it underneath the infra-red lamp to be re-introduced when the farrowing was complete. We never bothered again – sows are generally careful mothers, ensuring that all their piglets are in front of them, before lying down. We have lost one or two from being lain on, but Rick found this common too with the crating system on the commercial farm where he worked. It takes several days anyway for the weaners to discover the infra-red lamp for themselves and until then they will lie close to the warmth of their mother. It is best to keep an eye on progress until the afterbirth arrives; sometimes a piglet is born enmeshed in it, unable to breathe and requiring assistance. Once, after Grace had produced fourteen piglets, assuming all to be over, we retired to bed without awaiting the afterbirth. Next morning, we found seven further piglets dead. We could only suppose that she had got very tired and subsequently careless. Twenty-one is a large litter today and a sow, generally having between ten and fourteen teats, would find it difficult to feed them. I have not tried hand-rearing a piglet yet, but we kicked ourselves for our neglect. When keeping pigs as a supplementary income, it is hard not to think of each piglet in monetary terms. Walking in on a sow quite recently munching through a two week old piglet – it had probably been trodden on, pigs will clear up their dead – I groaned: 'She's just eaten £20'. Too

careful surveillance is not always helpful though – the sow might intentionally hold up the proceedings until spectators have moved away.

During one labour, Grace only produced a couple of young and then stopped. She pushed and pushed for two hours and finally seemed to give up. We called the vet out. Unsuccessful in performing an internal examination, he administered a sedative, pulled out a stillborn piglet coming awkwardly and holding things up, then gave a further injection to commence contractions again, which the sedative had arrested. Grace lay semi-conscious while the piglets seemed to be shed like peas from a pod. I am always amazed at how small the offspring are from such a large animal – but they certainly grow quickly.

We do not carry out tail docking, castration or teeth clipping, all of which are performed commercially. As they are born inside, we offer a clod of soil on a grass clump daily, from which the young extract sufficient iron not to warrant injecting artificially.

By about five weeks of age, the weaners are drawing a lot from the sow and tucking in well to meal too – the mill bills are a nightmare for these few weeks and so we try to sell them all off as soon as possible. There are several backyarders in the district who take stores from us at eight weeks, which is when we wean.

Pigs will definitely devour all the household scraps (except citrus peel). They will munch through all the garden waste, but the nutritional value of these scraps is low. Laws govern the feeding of swill containing meat, eggs and fish products. Potatoes, especially when cooked, are good for fattening, but twenty pounds is the equivalent of the five pounds daily meal allowance. To be at all competitive, we have now concluded that we must feed proprietary meal, with a balanced protein ration. It works out cheaper and is far less time-consuming.

Gallons of surplus skimmed milk to feed the pigs simply do not exist. With skimmed milk valued at fourteen pence a pint in 1982, such a costly way of feeding them would have led to expensive pork. It then cost thirty-seven pence a day to feed a pig with five pounds of sow and weaner meal. I cannot imagine a pig being happy with the equivalent in monetary value of skimmed milk – two and a half pints a day!

To start a breeding enterprise, however small, there must be money behind it to cope with all those appetites, including the increase in the lactating sows – two pounds of meal maintenance plus one pound per piglet, with an extra one pound maintenance if there are over eight young. This is where we have sometimes gone wrong, and it is false economy to skimp. Once, we built up to two sows with no float and ended up broke and selling them both, when weaned. The good thing with pigs is that they are quick to build up again, because whereas a cow or goat only gives birth once a year to one or maybe two offspring from about the age of two, the sow will rear litters of eight to fourteen at ten months old, and thereafter twice yearly (commercially for about six litters until worn out and retired to the meat processing factory). One has always to look at market trends – there is really no point in expanding when weaner or pork prices are rock bottom. So our pig enterprise has had to be adaptable. At the moment we are trying a different money-spinner – having bought in four landrace gilts at six months and put them in pig to be sold off as in-piggers a month before farrowing. We stand to make £30 per head, but our money has been tied up for four months, and it has been a worry with feed costs and the pessimistic thought of any possibility of a vet's bill to swallow the profit. This particular venture will not be repeated.

CHAPTER 13

They don't read books

In our idealistic dreams, honey would one day completely replace all sugar as a sweetener in our diet – what quantities this would entail! In 1979 we acquired two antiquated WBC hives – one second-hand and one in exchange for a rabbit. WBC hives are difficult to transport when occupied and probably would not have been our choice but we could not look a gift-horse in the mouth. So we ordered eight frames from a reputable firm to be sent by rail. The arranged time of arrival came – but no bees. On returning to the station the following morning, Rick knew the mislaid package had turned up before he opened the parcel office door – the angry high-pitched buzz of 16,000 bees must have been audible miles away!

That colony worked and multiplied at a tremendous rate. We took twenty-one pounds of honey from a hive not furnished until the end of May. For some reason which is not exactly clear now, we had been told to take the honey after dark. With much difficulty, tripping and stumbling this was accomplished, and then followed hours of spinning the honey out with a mechanical extractor, picked up second-hand for £15. What a state we got into, and that stickiness was transferred everywhere by our tacky shoes. This amount of honey was an untold luxury and we had endless licks until we could bear no more. The whole operation dragged on late, so we left the mess and went to bed ... and they swarmed.

Rick drove down the lane on his return from work at lunchtime and was greeted by a melodious buzzing; the air was thick with circling bees whilst I was obliviously engrossed in pulling strawberries. Rick donned his full outfit to take the swarm. Some beekeepers are so confident that they work the bees in shirt-sleeves and ungloved hands, with a veil as the only means of protection; but Rick will take no chances – he rather resembles a visitor from another planet as he tries to walk in his full outfit! I was commissioned to produce a white boilersuit from an old sheet for him to slip over his everyday clothes as a safeguard against the creatures crawling inside his clothes. I stitched a tube of sheeting, elasticated at the elbow to a pair of thick, white gardening gloves to seal off the cuffs; a veil and wellingtons and trousers tucked in, in the belief that bees always move upwards, completed the picture. The books warn that navy is to a bee, what red is to a bull, and that wool is also disliked.

The swarm had by now accumulated in a massive cluster around the queen on a shrub in the back garden, the bough bent to the ground under the weight. Rick gingerly snapped the branch, letting them fall into the box he had placed below. The fact that the bees seemed content to remain in this position, indicated that the queen was still amongst them. At this stage, they appear very passive - heavily drugged with the quantity of honey each bee has brought from the old hive; and in fact, swarming bees are not supposed to sting. An acquaintance wished to acquire a swarm, so he was informed and arrangements were made for its collection later that day, when flying had ceased. Evening came and all appeared still; Rick, fully adorned again, cautiously lifted up the box. Strange - it was as light as down. He raised the lid, to find just two bees left.

The next morning I had to summon Rick home from work - the bees must have returned to the hive the night before and now they were re-swarming. I hurried frantically up the lane to the call box and dashed back again, so as not to lose track of the drifting bees. They finally alighted on a ten foot copper beech tree on Garford's land. Rick sought permission to retrieve his swarm, which involved sawing off a branch and gave warning that it was necessary to leave the occupied box in position until dusk. He drove back to work for the remaining hour, but no sooner had he arrived home again, than Kath called out 'Did you know there is another swarm up this tree?' and sure enough, that enormous prime swarm had divided. Eric was duly summoned. He systematically went through the hive with Rick, destroying all the queen cells - a job that should be done weekly to prevent swarming. One of the swarms was settled into the second hive, but it failed to survive the winter; the other swarm Eric collected in the middle of the night, while we slept, in exchange for another home-made, second-hand hive.

The following spring, 1980, we explained to the bees that we were moving house, and we transferred them along with various tools and other essential outhouse belongings to their new situation, beneath the apple trees, facing south at Fir Tree Cottage.

Four days after we moved in, Rick and I were standing yards from the hives, when a bee attempted a landing on Rick. He flapped it away, in my direction, and it stung my nose. According to the book, a sting will come up for twenty-four hours, and will take another twenty-four hours to go down. My sting went on swelling for a week – my face looked so grossly contorted, my eye hardly discernible, that I wanted to hide away, and I did not feel totally fit either – I began to wonder if this was an allergic reaction.

The next time I was stung, some two years later, I was walking unconcernedly across the grassed path, mingled with clover, to my usual task of weeding and a bee busily working the flowers became entangled in my sandal; it panicked and stung me. Rick scraped the actual sting off – it should be done with a knife or finger nail, and not pulled out, which squeezes the poison sac, injecting more poison into the puncture site. I did not fuss over the wound unduly, and continued with my weeding.

When I woke up next morning, I was not sure if I were alive or not – my head ached, I felt sick, and more tired than when I had gone to bed. My foot was enormous, the swelling rose twelve inches up my leg. I could not move my toes or see my ankle. I must be allergic to bee stings, I thought, so I dashed down to the Saturday morning surgery. The GP seemed to find the whole incident very amusing. He did not think that I had an allergy, the unusual extent of the swelling was probably due to a secondary infection, from working along the ground without covering the injured area, he explained. I was advised to spend the day, foot up, but Rick's parents were coming up today for only their second visit in the two and a half years we had been here, and then the garden centre rang with a list of vegetables to be pulled as soon as possible. So I had to struggle on and smile and pretend everything was just fine.

Perhaps after a few stings, I will build up more immunity and react less violently. Rick finds that when he is stung, the inflammation occurs almost immediately and subsides completely within hours, with absolutely no adverse effects. Since my second episode, we have heard of several cases, medically confirmed as allergy to bee stings – nausea, vomiting, grossly swollen lips and respiratory difficulties. A friend experiencing the above symptoms, occurring within ten minutes of one solitary sting, was advised to be rid of his bees without delay.

In 1983 the hives got into a dreadful state. We had already filled the third hive with a swarm recovered from the gooseberry bush of a beekeeper in the village – and then our bees swarmed left, right and centre! The week our first caravan guests, Peggy and Eddie, were with us, we demonstrated how not to manage bees. One swarm settled on the privet hedge outside the caravan door, where it remained boxed for the length of their stay, awaiting to be reunited with the colony (achieved by destroying one of the queens); a couple more swarms were recovered from the Bramley trees overhanging the hives; when yet another swarm alighted on branches out of reach, Rick called for the hose-pipe; spraying water to imitate rain is in theory said to bring the bees down. What went wrong in practice I do not know. Perhaps the hosing had not been long enough and perhaps the resulting spray had not appeared down from the sky, but had driven in forcefully from the west; but the next thing I knew, the bees had taken off in the opposite direction and Rick was puffing off in pursuit.

When a boxed swarm re-swarmed, Rick admitted defeat and called in a member of the local Beekeepers' Association, a commercial beekeeper, to sort out the muddle. Pete arrived on a hot, humid afternoon. He smeared oil of cloves up his arms and on his hands, donned only a veil then, with thunder rumbling all around us, he opened the hives and, with the occasional puff of smoke from our smoker, which he scorned as no more than an ornament, methodically destroyed all the queen cells. The bees generally on the defensive under Rick's management, remained placid with Pete's calm approach.

The finale to the season's activities came when we sought advice from a chemical firm after the strawberry disaster later that summer – the representative on advising us to choose different crops to supplement our soft fruit income, beetroot and spring onions for example, suggested we should keep twenty hives of bees!

CHAPTER 14

Leaving the Marsh

At the end of 1979, after five growing seasons on the Marsh, we felt we had done all we had aimed to do on our limited acreage. The land was under control and running as smoothly and as intensively as was possible, and the challenge we needed was gone. We could not increase our livestock any further without becoming overstocked; all we could do to the house, within reason, had been done – and with three children, we were fast outgrowing a two bedroomed cottage. Sadly, my mother had died very suddenly back in the summer and had left me a legacy. We could have built an extension, but the problem was gaining more land, so, once more, we decided to move.

We gave the exterior walls of the cottage a coat of red-brick paint; the sills and drainpipes in contrasting cream. We added our personalised touch to the sterile-looking interior, which we had originally white-emulsioned throughout. The chocolate brown and mustard of the living room, the apple green in the bathroom, the pale green and cork tiled bedroom and the red, yellow, green and blue nursery (which gave the appearance of a big toy building block, with the sunshine rising out of the skirting board in the corner) all seemed more in line with our unconventional thinking. With Ivy Cottage prepared for the market, we went out to look for greater things in pastures new.

We soon realised that Lincolnshire was, without doubt, one of the cheapest areas in the country, the land the best, and decided to concentrate our efforts there. The day we made our decision, Rick was offered full-time employment for the winter, a godsend to us at that time, and Rick, of course, accepted, but asked if he could start on Tuesday. On Monday, we would go to the main towns of Lincolnshire and scour the estate agents with a tooth-comb. We planned to start at Boston (further up than this, property was more expensive) and from there work our way back south.

Boston pleasantly surprised us – here was a market town with history and charm; in its renowned buildings tucked away down side streets (the Guild Hall where the pilgrim fathers were imprisoned among them); in its splendid river and docks, once second only to London in importance; in its parish church, the 'Stump', the largest in the country; and in its broad, tree-lined main street 'Wide Bargate', where an extensive street auction selling everything is held each Wednesday.

It was in the second office we called at that we found it; 7.173 acres of grade one black land for £29,500. The house was not particularly appealing from the photograph, though it was functional for our purposes - three bedrooms, two living rooms, bathroom - but the land sold it to us. We drove out the ten miles northwards for inspection. Directed around the back roads, we arrived at a holding off the beaten track (later we were to find it just a mile from the A16 by a more direct route). We came anyway to change our ideas about living in the wilds; it is neither practical with petrol the price it is to be too far from the amenities (shops, post office, doctors, hospital, library, and so on) nor fair for the children to be cut off from the friends and outside interest they need as they grow up.

It was a detached house; the original part we assessed to be about a hundred years old, with an extension doubling its dimensions, in 1950s' style, but there was no one at home. We waited awhile, getting it all into perspective. The outbuildings were numerous - a fuel store and two small sheds, a greenhouse thirteen feet by seven feet, a timber and corrugated store, a timber and corrugated tractor shed, a corrugated bale store seventeen feet by ten feet, a double garage of timber construction twenty-four feet by eighteen feet, a timber garage fifteen feet by eight feet three inches, a brick workshop seventeen feet six inches by eleven feet three inches, two timber pigsties with runs, and a fine sectional timber building, formerly a poultry house, thirty-two feet by twelve feet. Apart from a few fruit trees and bushes close to the dwelling, the land lay fallow. There was open farmland all round. In the adjacent field a gang of agricultural workers picked brussels, but the nearest neighbour was 200 yards away. We knew we had to have Fir Tree Cottage, just as with Ivy Cottage before. We dashed back to consult our solicitor friend, who happened to be on a week's leave. A telephone call from him had the property 'under offer' there and then.

That evening we went back to view the interior with Jade, who had been at school. The old part was full of character; pine doors with wooden handles; a lean-to kitchen with sloping timber ceiling (my favourite room of the house); a huge walk-in pantry with meat hooks hanging from the ceiling; the living room with its built-in cupboards and shelves in the recesses; the master bedroom above, up the closed-in stairs. The work done twenty-five years ago had added a second sitting-room and extended the landing into almost a box room by itself, at the end of which were built two further bedrooms. The modernisation had also included a bathroom and separate WC facilities. Most rooms were on the southern side, thus benefiting from the sun during most of the day; and the kitchen situated at the back, would keep the setting pans of milk, cheese, flitches of bacon, and dried herbs cool the year through. This time round we had more idea of what to look for when buying an old property. We did notice where the damp course had failed in parts, that the wiring was antiquated, that the window frames were in danger of rotting from want of a coat of paint, that the roof spaces were uninsulated, that penetrating damp had ruined the wall-paper on one gable-end wall, but it made no difference. All these we overlooked, because we knew we were going to buy those seven acres before we even stepped foot inside the house.

We immediately put our own property on the market. It was valued at around £20,000 and two worrying months were to pass until a buyer finally turned up. The market was very slow at that time, 1979–80, made worse by the Christmas period, when no one cherishes thoughts of moving house. Our agent was reassuring that things would pick up in the spring, but there was pressure at the other end. Having been strung along for three months by a previous buyer, Mr Wilson was anxious not to be let down a second time. All we had been able to do was put down a £100 holding deposit with his estate agents; a bridging loan was out of the question for us.

Then, one afternoon in February, a gentleman drove into the yard, and not only was he really keen on the place and searching for his few square yards of the earth's surface (it had been our hope Ivy Cottage would fall into the hands of folk who would carry on from where we had left off) but his family were desperate for accommodation, having already sold their own home. Within a few hours, we heard that a formal offer of £20,000 had been made for Ivy Cottage and we immediately accepted. Once contracts had been signed for a completion date of 21 March it was just a case of tying up the ends of what we had started months previously – packing up, taking up fencing, and generally tidying. Mr Wilson allowed us to put in our herbs at Fir Tree Cottage, and to mark out the land into acre and half acre plots, but he was less happy when we turned up with the bees. We transferred all the outbuilding stock – tools, netting, stakes to an empty shed. It was only thirty-six miles and with the 1963 Landrover we now owned and Eric's trailer, we could take a large load at a time.

For the big day though, we hired a removal van. There was a price-cutting war between the local firms and we did well out of it. We chose the lowest estimate, given without an inspection, and I feel they rather got the raw end of the deal. Even after our stints each weekend, we had far more than the usual two-bedroomed cottage – cream separators, grain grinder, sacks of household grain, solid furniture, all of which were very heavy. The vehicle was dangerously full when the tailboard was raised, and the Landrover took the overspill. Even so, we had to leave behind the cast-iron mangle we had intended as a cider press – the last we heard the new people were converting it to extract oil from sunflowers. Then, at the last moment, as Audrey waved us off, we felt torn apart; all at once we had mixed feelings about leaving the Marsh for the big new venture.

PART II

CHAPTER 15

And this is the twentieth century

We were transposed into a wet world. For days after we moved, rain descended without let-up. Two big puddles appeared in the field to the west of the house. Even after the downpours had abated and the birds burst into song again, it was some time before we could get onto the land. We were impatient to drill our seeds - grass, grain and animal fodder - to plant the vegetable garden and establish the tree nursery and soft fruit area with the stock we had brought with us, so when the surplus water had been motivated by channels dug to the dyke, Rick drove the tractor too eagerly onto the still tacky soil. To this day he has not lived down the reputation of being 'that idiot who has moved down Scarborough Bank'. Harold Panton, our friendly neighbourhood farmer, was out for the morning, so the tractor remained in its stranded position for all to see until lunchtime, when he was able to get a tow.

The tractor, a David Brown 25D, had been bought from Mr Wilson, along with various other equipment he would no longer have need for in a bungalow in town. There were the tractor implements - a two furrow plough, spring tine harrows, drag harrows (duck foot and spiked) potato ridger, row cultivator and a snow plough - also a Scotch cart on rubbers, a pig trailer, (which we could not do without) potato trays, potato picking baskets (that serve many purposes), all for £300. The tractor admittedly was in need of a new clutch - and still is - and one of these days we will pick up a Cambridge Roll, a set of disc harrows and a muck spreader to complete our independence.

We had really come to a back of beyond sort of place. The paper man left our weekly local at the house at the end of the bank (half a mile away) and the milkman called down the road once a week with sterilised milk. We made our acquaintance with the immediate community; Grace next door with her husband, Nev, employed as a gardener, and son, Colin, an athletic-type; and on

the far side, Low and Mrs Baldock, a couple in their seventies, the parents of Mr Wilson's housekeeper – the same old story of half the village interrelated! Former smallholders, they had sold the main part of their land, but now long past retirement Low still holds down a full-time job at Harold's poultry farm. Further up Scarborough Bank, there were the Pantons, another large local family, the Dafts and Charlie Rymer and his wife, all making a living from the land.

Spring arrived and all around us the earth was bursting into life, the roadside grass, the budding hedgerows and our crops all growing. April proceeded, as it always has done, with the birth of goat kids, a flood of eggs from even the oldest hen, full production from the rabbits, the bees flying out on fine days for the earliest flowers, and the mating season of the geese well under way. With our increased acreage, our stock increased accordingly. Blossom, the house cow, arrived to graze the roadside verges until our freshly sown pasture was established enough; and we had five cade lambs to hand-rear – the commencement of our flock. Our holding was registered with the Ministry which meant we could use their advisory service free of charge, together with the publications, including a monthly newsletter.

In the immediate area there were a number of families practising some form of self-reliance on a small acreage in a variety of ways and it was the same situation throughout the county – families were heading in their hundreds to Lincolnshire, where property with land was still to be had relatively cheaply.

Winter came to complete our first year's cycle here at Fir Tree Cottage; December, the month it is impossible to dry the clothes – when the days are so short, it is nearly dusk before the wash is even done. Then the frosts only harden the motionless garments or if the wind does blow, the pegs do not hold them to the line. It is a battle just getting the wet washing near to the stove, competing with huddled children, and when one load is at last aired, the laundry basket is spilling over again. A season spent with mud as our carpet – the soil peat on clay is tackier by far than our previous silt on sand (and dust is the problem in the summer); of daylight hours so brief that our pace slows to such an extent it is hard to start at all. Then spring returns, days lengthen, the earth comes to life and with it, we renew our energies and by midsummer we are working outside for fourteen hours in every twenty-four.

This is really living life by the seasons.

We tried everything in our initial years here - our fanaticism left too little room for pleasure. In trying to do everything - grain growing, haymaking, stockfeed, a large area set down to labour intensive soft fruit - there was no time to 'stop and stare', and the enjoyment was lost, because the above tasks that should have been fun, always had to be a race against time, as other jobs were pressing. But we can at least say we did it all.

It is three years since we moved to Stickford and still the decorating has hardly been started. We stretched our resources to the limit in buying the place and so far any spare cash has been spent on the land. We never seem to have time and money both together. If we have an unexpectedly wet week with nothing else pending, the chances are that the cash flow is at a standstill. In the summer, the only time money does flow our way, we are much too busy to even contemplate decorating. Then when we do have the time to breathe again, other things swallow up the spare cash first - more trees, fencing stakes, repairs to buildings.

Although we do like to collect nice things to look at around us, beautiful, hand-made objects, even better if we have made them ourselves, the overall appearance is of a basic peasant dwelling. In the main room for example, one carpet was obtained for thirty-five pence at an auction, the other and a rug for free. The dresser and the heavy bookcase inherited in former properties were stripped and restored for nothing. My favourite armchair was retrieved before it was consigned to the tip - only the dining table and chairs were actually bought with hard cash, though at a realistic junk shop price, but to our eyes at least the overall effect is appealing and the atmosphere is comfortable and a pleasure to come home to.

When Rick's sister, Sue, was summoned to our aid to run the home, while I had a short spell in hospital, she had no vacuum cleaner, no refrigerator, no immersion heater, no television; just a radio jammed on Radio Four, an antiquated twin tub and the choice of a calor gas cooker that takes all day to cook a joint of meat and runs out at the least appropriate moment, or of running the Rayburn in September, on a continuous supply of wood, that first needed chopping. There was also no clock whatsoever - and this is the twentieth century.

When our consumer goods do wear out - the washing machine,

the freezer, the cooker – with our peasant economy it would be out of the question to replace them with new appliances, so we buy second-hand from private advertisers. We find it is even cheaper to buy another second-hand washing machine or whatever than pay to have it repaired. I simply loathe mechanised devices – if it is not the household appliances breaking down, it is the car, the tractor, the Rotavator or the lawnmower. Rick does not enjoy mechanics either, and we prefer to earn money in some other way in order to pay someone who does to fix it for us.

The same policy applies to vehicles as the other mechanised equipment. We find it preferable to spend a couple of hundred pounds on a vehicle that we can run into the ground. There is a vast breaker's yard locally, where scrap engines and other parts can be had very cheaply. We had a Landrover when we moved, which we liked to think suited our life-style, but we could not afford such stylish transport – the eighteen miles to the gallon was horrific and spares were terribly expensive. The van was more in our line; one felt such a sense of speed as it rattled and shook along at forty mph. When we bought our present 'L' reg. car, relying on the tow bar to increase its capacity, we thought we had gone up in the world, until we realised the van had been a 'K' bought two years ago.

But although funds are usually low, we never get into debt. If we cannot pay for something, we go without it and I think that it is the only way we can live as we do. It is bad enough just paying the general bills sometimes, without the headache of second mortgages, bank loans and HP repayments.

CHAPTER 16

Cows and calves

Goats had been good to learn on; ideal for a small acreage, particularly in our situation with more free roadside grazing than we could ever utilise, but at Stickford we immediately broadcast two acres of permanent pasture, with the intention of keeping a house cow and sheep.

We bought Blossom Dearie, a down-calving heifer, but as the grass was not yet fit to be grazed, we tethered her along the verges of Scarborough Bank. Cows lack the goat's agility to graze the steep dykeside, but even so the chain allowed her access to a wide arc and we would move her stake several times daily. The weather conditions being favourable, within six weeks of sowing, Blossom was able to be brought into the field, divided and fenced by pig-netting into two one acre paddocks. Although this was only three feet high, Blossom never showed any interest in discovering what went on outside her boundaries.

Blossom calved a month later than the date of confinement we had been given. Her bagged-up udder at least was reassurance that something would happen one day, and at the end of June it did (not the right month at all for a calving - missing the advantages of the spring-September flush and consequently with the calf price sky high in the market if the cow is to be a suckler). Blossom had been artificially inseminated to a Murray Gray, which produces a small calf with a rapid growth rate.

At long last, she took herself off into the furthest corner of the field and, in pouring rain, gave birth to a heifer without any problems. We could not wait to start hard cheese-making, so we sold the calf on to a relief milker/smallholder in the village who intended to raise her up for a house cow. In fact, we could have made cheese and suckled the calf with Blossom's milk yield of twenty-six pints a day.

One should aim for a calf per year from a cow and, as we wanted to bring the event gradually back into the spring, when she showed signs of bulling sixty days after calving we rang the AI centre. Here one of the advantages of the cow over goats became apparent - no humping a noisy goat into the back of the car and travelling miles in the night to a smelly billy goat - we just picked up the telephone before 10 a.m. on the day, gave name, address and the bull wanted to a recorded message and then went about our everyday business. In due course the AI man arrived, donned protective clothing and wellingtons, waded to the cow shed, inserted the straw of semen into the haltered animal, disrobed, cleaned up, the AI certificate and fee changed hands (only £5.29, £4.02 for a repeat). No longer than ten minutes elapsed and he was away to his next call - a very busy man like our vet.

The fact that Blossom was a Jersey cow posed a problem as to what breed to AI to. A Jersey heifer was worth around £50, but bull calves very much less. Cross-breeding produces a better beef proposition; though not as valuable as a heifer kept pure. It seemed foolish to use a large breed and commonsense to put a heifer to Jersey, Angus (prime beef but with a slow growth rate) or Murray Gray.

We chose the Jersey bull of the day. A couple of days after being artificially inseminated, we spotted a bloody discharge on her flank - confirmation that she had actually been on heat. After the gestation period of nine months, Blossom gave birth to a bull calf - at the beginning of June - with some assistance. It was a simple breach, pulled out by rope without much effort. We had not been very successful with our cheeses and so this second lactation we decided to buy in Jersey calves to rear, and use some of the money we earned to buy in cheese.

One of the acre paddocks had been set aside for hay, and Blossom and the few sheep made little impression on the other lush field. When the haymaking was over and the paddocks were both in bounds again, the grass grew too quickly for them. The question of under or overgrazing is a difficult one. The owner of six ewes with newborn lambs at foot coping with a field in spring flush may find the same field hopelessly overgrazed later (July and August) as the lambs grow and the grass slows down. We needed something to eat the grass quickly before it went too far so we bought a second down-calver (to cover Blossom's dry period, to suckle during the remaining months). Our intention had been to bring back a heifer, just in calf, for approximately £200, but when we got right down to Suffolk in our old van with borrowed trailer, the choice seemed somewhat diminished from what it had been over the telephone three hours previously. Cello was three months off calving and priced at £275. She had been overwintered outside with the herd, which might have been why we had such a job getting her into the trailer.

The head gasket blew on the long return journey, and the last forty miles were covered at snail's pace. It was midnight when we drew into Stickney; we had had nothing to eat since midday, but the children were passed caring anyway, fast asleep. We just left Cello in the trailer and went to bed, intending to unload her first thing in the morning. We got up at 6 a.m., our eyes still heavy with sleep from the exhaustion of the troubles of the night before. But they were not over yet. The tailboard was lowered, halter slipped over the cow's head, and ... nothing happened. Rick yanked on the rope, but she did not budge. I joined in tugging and Jade got behind and shoved, and slowly she sauntered down the slope. Then she stopped again. The procedure was repeated, raising our blood pressure and tempers no end and no doubt not doing my hernia a lot of good either! We walloped her flanks, raised her tail and tried all the other tricks of the trade, all to no avail. We had been shouting and swearing for half an hour and she had barely moved five yards!

Jade was sent to summon help, but even with extra muscle power, she remained rooted. The half an hour of exertion slipped by into an hour wasted, when someone suddenly had the brain-wave of bringing Blossom down into the yard too - the sight of another cow might do the trick. Rick led Blossom down, round in a circle under Cello's nose and began walking her back again in the direction of the paddock. Without warning, Cello charged. Her feet kicked behind her like a stampeding buffalo; but instead of clouds of dust, squashed strawberries, flying leaves and trusses went in her wake, as she took a short cut across the soft fruit block after Blossom! Rick closed the gate behind the two Jerseys, got in the car and, better late than never, accelerated in the direction of the pig farm, where he should have been at work.

Cello remained timid, wanting no human contact; a month later, when Rick's sister managed to touch her muzzle with the tip of her finger, it was quite an achievement. When her confinement drew nearer, Rick began at milking time to offer a small amount of concentrate, while Blossom enjoyed her ration. As she became used

PW

to taking the nuts from some distance away, Rick would move the bucket closer and closer to his side. In time she came right up to him, but should he make any move to actually touch her, she was away and nothing would entice her back again that day.

October came round; dampness lingered in the air and there was a touch of frost some mornings. The cows were brought down before we put the clocks back, and once shut up in the shed, we felt happier in the knowledge that we could at least supervise Cello's calving here.

One day, Cello was obviously in the throes of contractions before Rick reached home after tree felling, but he was in time to see the appearance of the first foot and that was it. Nothing more happened. The calf was correctly positioned, but it was a good size and needed a tug. It took some effort from both of us before that Murray Gray cross slithered into this world. It was a heifer and we named her Sycamore after the wood Rick had been collecting that day. Cello's milk was not yet needed for the house, so we permitted her to suckle her own calf and bought in a second calf, a Hereford × Friesian, for her to accept alongside it.

As Blossom's second lactation petered to a close, we weaned the latest calf off Cello and prepared to milk her. Rick placed her concentrate beside Blossom's, as was now the usual procedure and settled down to begin milking – Cello stiffened . . . and was away. Rick jingled the bucket of nuts, but she carried on. It became a chase, Rick with lasso after a stampeding Cello. Rick panted and cursed and bellowed for me to come and help. By now we had extended the grass area to four acres, which only doubled the problem! Blossom was patiently tied up waiting to be milked looking with an air of disapproval at the spectacle. This was always the case when one or other of her kind was playing up – she would stand back like an elderly lady tut-tutting, never once was she the cause of the commotion and never would she join in. We finally had to lead poor old Blossom down to the winter cows' quarters in order for Cello to follow suit, where she could easily be caught and haltered, and this had to be the twice daily procedure from then on, even when Blossom had a calf at foot. It was not an infallible method however. Quite often if she had the devil in her, she would miss the doorway into the shed altogether and head off in another direction.

Cello's yield came nowhere up to Blossom's standard; at six

weeks prior to calving she was not keen to dry off, giving eight or ten pints a day still then. We would milk once a day for a while and then once every three milkings until she ceased producing altogether, but it was not long before her udder bagged up in readiness for her third calf, filling with colostrum, the special first milk containing antibiotics. One day, she once again took herself off into that same far corner of the pasture and soon gave birth with ease this time. It was a pure bred heifer and as the sun shone down and dried it off we called her Sunshine, and that same sun continued to shine for many days after, as the bundle of bones trotted after her mother on her gangly unsure legs.

Friday being four days following parturition, I milked Blossom right out, going by the letter of the book, which stated that to guard against milk fever, not to milk out completely until the fourth day, so we thought we were doing everything right. Rick went up to Blossom with the milking bucket first thing on Saturday morning and Blossom would not get up. She was just sitting there and her eyes looked strangely dull. There was obviously something wrong and suspicious of milk fever, because of the stage in her lactation, the vet was summoned very urgently. We propped Blossom up as instructed – it being vitally important she should not actually 'go down'. After what seemed an eternity, a car's brakes ground to a halt in the yard, the vet raced up the pasture and after taking her temperature and examining her eyes, confirmed our diagnosis of milk fever. Fortunately we had caught her in the very early stages – the vet pumped one bottle of calcium directly into her jugular vein to take immediate effect and a second subcutaneously. After one attack, we were told, a recurrence is almost certain in subsequent lactations, but to lessen this possibility, an injection to be administered a few days prior to calving is available. Blossom should not be milked for a day or two – Sunshine would take an adequate amount to ward off mastitis. How typical – having sold Cello only the night previously, we would now have a milk shortage in the house! After about an hour, Blossom stumbled shakily onto her legs. By Monday we were milking normally again, but it took some time before her yield picked up, not helped by the fact that on Tuesday, upon straining the morning's milk, clots appeared on the muslin – mastitis. Rick went into town to fetch antibiotic pencils to shoot up the affected quarters, which rendered half her milk unsuitable for human consumption for four days. The vet commented that this had been pretty inevitable – probably caused by bruising when Blossom had gone down heavily onto one side. Besides injury to the udder, other causes of mastitis are lack of hygiene and the spread of infection by flies. It can be detected by a distinctive salty taste to the milk. As a precaution against mastitis when drying the cow off, it is best to shoot 'Dry Cow', a long-acting antibiotic up each quarter.

There are tight regulations regarding the sale of milk and certain standards must be met before a licence is awarded. The Ministry come round to perform blood tests on all cattle over twelve months for brucellosis annually, and a skin test for the detection of tuberculosis every three years. Herds that are cleared become accredited. When Blossom was our sole head of cattle, it did sound rather grand to receive a document stating we were now an accredited herd. Licences are unnecessary for the movement of cattle within accredited areas, but a record must be kept. Cows should be wormed at regular intervals and a salt lick be available at all times. As with immunisation, prevention is better than cure, so it is advisable to supply extra magnesium, in the form of pellets, with the concentrate ration, made palatable by the addition of molasses, from when the cows are put out to grass up until July to guard against grass staggers. When first turning the cows out again in the spring, it is safer to limit their intake of the new lush pasture (and their grazing of kale and other succulent wet crops too), at the same time keeping a close observation until the appearance of loose droppings indicates that the grass is passing through. Bloat happens when fermentation takes place within the rumen, which is discernible above the back bone; then the gas has to be rapidly expelled, either with a defoaming agent (corn oil, sunflower oil, soybean oil, peanut oil or similar) or in the case of an emergency when the animal is weak and staggering – the dramatic method used in Thomas Hardy's *Far from the Madding Crowd* has to be resorted to – piercing the rumen for immediate release.

We had thought that in getting rid of Cello, our life would be quieter and more predictable. Perhaps it was for a while, until Mulberry, a bought in calf, grew to adolescence and began to have regular heat periods. Then for one day every three weeks, we knew nothing would contain her. She would step over the three feet six inches high door of her pen, if imprisoned in the cow shed, within minutes blundering her way through the outer door. There was hardly any point in returning her to behind the paddock fence, because as soon as the fancy took her, she would be running wild again. To make matters worse, Mulberry kept returning after service, so month after month, we would have this disturbance. After the second repeat, we sought veterinary advice. She was artificially inseminated for the fourth time around, then within an hour the vet had arrived to check her out - there were no genital defects and she was not a freemartin (ninety per cent of heifers born twins to bulls have incomplete reproductive systems due to interference by male hormones during development). After a hormone injection, we were advised to have her artificially inseminated on the second day of her heat period as well, to be on the safe side, so Mulberry was successfully AI'd at the fifth attempt. As she is to be our second house cow, we have left Blossom unmated for two months to avoid a milkless period. After all these service fees and a vet's bill, it will be a very expensive bull calf indeed!

DAIRYING

All the books I have come across make the dairying side seem off-puttingly complex - I hope I have put this across in the plain straightforward manner in which it would have been performed by the simple kitchen maid.

BEESLINGS PUDDING — The cow's first milk after each calving is colostrum, rich in antibodies. It takes four days to completely pass and the milk, as we know it, to come in. This colostrum has a special property in that it solidifies on heating. Many is the time we have boiled milk for a bedtime drink a bit too soon and not been able to pour it out of the saucepan! However, basins of the thick yellow substance put into a slow oven with a tablespoonful of sugar and a sprinkling of nutmeg will make a delicious egg custard.

CREAM — After each milking, we strain the milk into a settling pan and leave for the cream to rise for upwards of twelve hours. It is then skimmed by hand. We do have an Alfa Laval cream separator, bought second-hand when we had goats, but cow's milk is so easy to skim, the fraction more cream extracted mechanically is not worth all the extra washing-up of the numerous cones afterwards. We eat delicious cream on everything in generous quantities - whipped up occasionally if we feel the visual effect is worth the effort.

SKIMMED MILK — This is the milk we use for the house, so that we have adequate quantities of cream left for butter making. The hand-skimmed milk remaining is in any case probably as rich in cream as the Friesian bottled sort!

BUTTER — The cream is amassed in the freezer, until we have a reasonable quantity for churning in our 'end over end'. It is alleged to take the cream from eighteen pints of milk to make one pound of butter. The temperature at which butter 'comes' is 60°F. Below this it can take an awfully long time and so we heat it up. When the cows have been brought inside for the winter, the butter is paler, crumbly and unspreadable. Butter produced on grass is creamy and looks as if nasty colourings have been added! The butter is then washed, salted to taste and the finished product stored in the freezer until needed - it is used in all our baking.

CLOTTED CREAM — We do not get around to this often - partly because it is so extravagant on milk and partly because we have unlimited quantities of fresh cream anyway. Leave the milk from one milking for twelve hours as usual, then heat slowly until a skin forms. Leave for twenty-four hours. The cream that is then skimmed off is that glorious cream of Devonshire teas, and will keep for about a week. Our first summer with Blossom, we would pile our bread at breakfast with strawberry jam and thick layers of this clotted cream until we felt sick.

YOGHURT — Heat the milk to 180°F (pasteurisation point); allow to cool to somewhere between 100–120°F; stir in half a cup of live yoghurt. Pour into a vacuum flask (we have tried airing cupboards, the shelf above the Esse and other warm places, but this is the most reliable). Leave for twelve hours before checking. Flavour according to taste - fruit, nuts, honey; it is delicious with muesli or fruit crumbles. Keep back some to use as a starter for the next batch.

ICE CREAM — This simple but heavenly recipe is really only suitable for those with a house cow, who do not think of their ingredients in monetary terms! It is so quick and easy, no messing around heating up eggs or anything. Dissolve sugar to taste (but preferably honey) with a minute quantity of water; just enough to dissolve it and if using cocoa or coffee for flavourings, add these powders at this stage. Cool. Then stir in all the fresh cream you have available (I take two milkings, say about two pints of cream) together with soft fruit, nuts, alcohol or essences if desired. Place into the icebox of the refrigerator or the freezer and stir thoroughly about every thirty minutes until solid, to prevent ice crystals forming. Never will you buy those cheap imitations again!

SOFT CHEESE — We found out early on that it is no use leaving the milk to go sour in the passage of time, even in the heat of the greenhouse, it will just as likely go green first! So a curdling agent must be added. Rennet forms too rubbery a curd for soft cheese-making; vinegar works fine, but it tastes, therefore it can only be utilised where the flavour is masked with the addition of herbs or shallots. I like lactic acid best. It is fairly expensive at ninety-five pence for 100ml; a teaspoon will turn one pint. Then a tablespoon or two of the soft cheese could probably be used as a starter for the next batch, though I havc not tricd it. The curdled milk is hung up to drain through muslin. It has a bland flavour, so is ideal for cheesecakes and its soft texture is far superior as a spread to the other methods described above. Cream can be used for a more extravagant and luxurious lactic cheese with this recipe.

Basic Cheesecake

Base
4oz butter
8oz digestive biscuits (crushed)

Topping
1lb soft cheese
8oz sugar
2 eggs

Method Melt butter. Stir in biscuits. Press into bottom of cake tin. Mix together ingredients for the topping. Pour onto the base. Bake at 350°F for approximately 35–40 minutes. Decorate with soft fruit and whipped cream if desired.

Oriole's Lemon Cheesecake

Pastry base
1lb cheese (soft)
4oz sugar
1 lemon (grated rind and juice)
3 eggs
2oz sultanas
grated nutmeg

Method Line a tin with pastry. Beat the eggs, cheese, sugar, lemon and sultanas until smooth. Pour into base. Grate on nutmeg. Bake in a moderate oven for 40 minutes.

Curd Cake

2oz margarine
3oz sugar
2 eggs
8oz cheese (soft)
2oz flour (with baking powder)
1oz chopped dates
1 tablespoon icing sugar

Method Grease and line tin. Cream margarine, add sugar. Beat in eggs, then cheese. Fold in flour and dates. bake for 40 minutes at 375°F. Turn out when cold and dust with icing sugar.

CHEESE-MAKING

When Blossom calved for the first time, we could not wait to be rid of her calf so we could start hard cheese-making. I enthusiastically turned out cheese after cheese, once every four days (the amount our press will cope with) for the six weeks the recipe stated was the minimum time necessary that the cheese should be left to mature. On the alloted day we eagerly tried the first cheese - it was so absolutely foul even the pig would not eat it! This rather put me off cheese-making and it was at least eighteen months before I tried again.

At the time, we were milking both Blossom and Cello - Blossom due to be dried off and Cello's latest calf had been sold, ready for her to supply the milk for the house. We tasted the first effort after only two weeks, and it was very good cheese indeed; so cheese-making became part of our everyday routine, along with the butter and bread-making. We can not account for this difference, but here is how we do it.

1 We take the skimmed milk from two milkings, unmeasured and not using a starter, that seems too much like a scientific experiment! I prefer to put myself in the place of a farmer's wife, maybe at the turn of the century. However we have used 'shop bought' cheese, quite unnecessarily, as a starter before; but find the bacteria within our muslin and followers now act as a starter to give a consistent flavour to our cheeses.
2 Heat the milk to approximately 80°F.
3 Add the rennet - one teaspoon to one gallon.
4 After about half an hour, or when the milk has set, cut into small cubes and place the pan over a very slow heat until the whey has risen.
5 Remove from heat, then pitch the curds, that is, leave in the whey until the correct acidity is achieved. To test for this, hold a piece of curd against a hot iron. When threads form half an inch long, this stage of acidity has been reached. I take this now to be from breakfast until teatime, but it is a good idea to check occasionally - conditions can alter, for example milk sours more quickly if the weather is warmer.
6 Drain off the whey for the pigs. Break the curds up into tiny pieces. Put these into the mould, lined with muslin, and press.

After half a day, turn the cheese over without removing the muslin and re-press. When pressure has been exerted for twenty-four hours, renew the muslin and turn the cheese. Repeat until the fourth day when, if the whey has finished draining, remove from the mould.

7 Store on wood at 50–60°F away from mice, but where the air can circulate; something like a slatted clothes airer, suspended from the ceiling would be ideal. In the summer months paint with cheese wax or lard, to protect from blowflies.

CHAPTER 17
Little and often

After our disastrous attempts at hard cheese-making during Blossom's first lactation, we decided that when she calved again we would venture into the calf-rearing field. Then with all the money we would make we could buy in as many cheeses as we desired. Ignoring such advice as 'I have lost more money at calf-raising than at anything else I have tried' from a long-term smallholder, we started our enquiries. I tackled the vet, here for the routine yearly brucellosis test, as to where we could purchase drop-calves with the view to avoiding the markets - having read that such places were arenas for the spread of infection and the cause of much stress to the new-born animal, which incidentally must be a minimum of seven days old by law before entering. He explained that in this area, dairy farmers mostly calve in the autumn to ensure the winter milk cheque; thus using the spring and early summer grass to boost the yield. Therefore, few calves were in the market from May to September and these demanded top prices. Blossom was due on 6 June.

This changed our course somewhat. We would buy in Jersey calves and raise them to sell as down-calvers. There was a commercial herd of six gallons a day milkers, two and a half hours drive away, that calved throughout the year. At this time, a drop-calf cost £40 and a down-calving heifer was worth about £300. So, as soon as Blossom had given birth in the pouring rain to Cobalt, a Jersey bull, we arranged by telephone to pick up a Jersey heifer calf (price £40) and a Jersey × Charolais heifer calf (price £30) that we intended to raise to sell for meat.

It was a pleasant drive there on a hot summer's afternoon, the window down and in holiday mood. Once off the main road, we entered beautiful countryside - high hedgerows, lush grass fields, narrow winding lanes. We found the farm - a large, long, white house, built in the tradition of the region - alongside an expanse of attractive old buildings and a milking parlour, surrounding the yard. Mature trees were liberally dotted about and a village pond opposite completed the picturesque scene. Around the corner of the road, plodding slowly along, came the milking cows, their distended udders swinging, the cowherd bringing up the rear in his fisherman's smock, moving on their twice daily journey to the milking parlour, at 5 a.m. and 2 p.m. Progress was steady, each taking up her particular place in the routine proceedings with the ease acquired by habit. One, freshly calved, lowed deeply and softly to her young one in the adjacent byre and a friendly nudge was necessary to keep her moving. When the first ones had been installed in position, the herdsman was able to continue alone and Mr Morris joined us.

We were shown our two calves; the pure Jersey who came to be called Mulberry, was four days old; the cross-bred, named Majollica, only two days. Mr Morris let us look about, much to Clyde's delight, as cows are his obsession. He explained that in this sheltered area, he was generally able to leave his stock out until January. We looked at the enclosure of calves of varying ages. None were put out to grass for the first year. We were told the heifers were all AI'd to Aberdeen Angus to produce a small calf, rather than to a Jersey, as Jersey bulls are relatively useless; all were sold as drop-calves to a veal producer for a couple of pounds and the heifers were not retained from an unproven cow. Older cows from which he did not wish to take a heifer were put to a Charolais or Murray Gray. Feeding concentrate, he had these to slaughter weight by thirteen months. It was a long haul back, so we paid our money, collected the AI certificates and fitted Mulberry and Majollica in the back of the van.

Back home, we settled the new calves together with Cobalt in a pen and immediately fed them a drink of milk after their hot ride (probably our first mistake). We had bought three plastic buckets and were splattered as we attempted to teach them the art of taking refreshment with head forced downwards; whereas all the goodness they had ever known in their short existences had come from their mother's udder up above. They made slow progress; gradually mastering the art by sucking milk from our fingertips. Majollica was the last to understand the technique. For days she barged and pushed in her frustration; although she stood inches above the two Jerseys, she represented brute force and ignorance. At length, she too demolished her milk ration in a few noisy gulps. Mr Morris had told us to commence at two pints daily and gradually increase to a gallon. He himself only fed once a day, which seemed dreadful to us. We read our books; the house cow textbook written from an American angle and the Jersey Cattle Society booklet we were sent free of charge. Opinions differed, so we settled for three feeds a day.

It was when the calves were taking five pints a day, that things seemed amiss. Majollica developed a slightly wobbly gait; she would stumble if her companions knocked against her, and no longer bullied us, as we held her bucket. Then we noticed her scour – a foul-smelling, light-coloured diarrhoea, the dreaded white scour we had been warned against. We rang the vet to arrange an appointment to bring her to the surgery, but were told it was their policy to visit calves on home premises. So in due course the vet arrived, took her temperature via the rectum and felt in her mouth. He explained that if the mouth was cold, there was generally no hope, but Majollica's was still warm. We were given antibiotic tablets and a three-day course of Ion-aids – half glucose, half salts to be fed in water, instead of milk. We carefully obeyed the instructions for the prescribed period and the scours seemed to cease, so we began to reintroduce the milk as the vet had suggested – a quarter part milk to three-quarter parts water; half to a half; three-quarters milk to a quarter water. But after only a day back on milk, poor Majollica recommenced scouring. Over the phone, the vet ordered us to stop the milk feeds again and to give water with added glucose and table salt until the diarrhoea was clear and then try again. But, as soon as we reintroduced the milk once more, the terrible stench confirmed that all was not well. It was Sunday when we telephoned him again in desperation and had to drive out to his house in the hills for more antibiotics. He then suggested that she could be allergic to milk. Next morning not only did Majollica seem far worse, reluctant to rise at all – but Cobalt too was scouring.

Rick decided to consult another veterinary practice. He explained the situation to a large group some distance away, who were highly recommended by several large stock farmers in the vicinity, but noted for being more expensive. They agreed to examine both calves at the surgery, but first stipulated our former practice must be notified that we were seeking a second opinion. The vet took a rectal swab from Cobalt to be incubated in their laboratory, but as Majollica had already been dosed with antibiotics, the results of a swab taken from her would be inaccurate. Both were given an injection of a long-acting antibiotic and treatment was advised. He requested us to ring with progress reports, so that he could adjust the therapy as necessary. Cobalt responded immediately, but Majollica faded slowly away. The knackermen collected her body, just bones held together by a thin, balding skin - a scouring calf loses hair firstly around the nose, then the ears and elsewhere. We were duly sent £1 for her remains.

Our sad initiation into calf-rearing. It sounds heartless to say we learned by our mistakes. The Ministry leaflets give alarming statistics for calf losses; but the next nine bought-in calves to date have all been raised successfully, and those that have scoured, have recovered. Majollica was only two days old when she was sold to us, too young for the stress of such a long journey. In no other field have we found more conflicting advice than with calves; there are many different schools of thought within the veterinary circle itself, as to the treatment of the scours - we have used three different practices and heard the opinions of several people within those practices. Now, we basically administer a shot of long-acting antibiotic, withdraw all milk and crush Tribrissen (Trimethoprim and Sulphadiazine) into glucose and salt water twice daily, until the course of tablets is gone. Pneumonia is another danger to watch out for, brought on by badly ventilated sheds or, alternatively, by draughty pens.

Opinions as to how the calves should be cared for to guard against scours are also widely varied. The advantages of bottle feeding are that calves can not drink it down too quickly, and that it imitates sucking from the udder. On the other hand, bucket feeding is less time consuming, and buckets are easier to keep clean. Feeding little and often again imitates how the calf sucks from its mother. However, others advocate feeding once or maybe twice a day, as with multiple feedings a ball of curd forms in the stomach and does not have time to be digested before further curd is added to it. Some say that milk should be fed warm, to resemble milk sucked from a calf's mother, but with automatic feeders milk can be fed cold. Putting a calf directly into the field is cleaner and more hygienic. A calf with its mother is protected by her from the elements. 'Born in the field, survive in the field', says Gordon. With an orphaned calf, you are its mother and so must ensure that it has shelter from rain and wind. Others say, however, that calves should not be put out to grass the first summer, as they readily pick up worms.

Anyway, Cobalt and Mulberry after disbudding and only eight weeks on milk at six pints a day, plus one bag of nuts, spent the summer on freshly sown pasture, ahead of the adult cattle, so that any worm infestation was an impossibility. After bringing Mulberry down for the winter, Cobalt was slaughtered for the freezer – 130 lb of beef for virtually nothing. There was no bright yellow fat, in fact, hardly any fat at all – the flesh was pale and the taste superb; perhaps it would have been classed as veal. Mulberry, in her second summer on grass, is now a fine heifer – the AI man commented extremely favourably on her when he attended her the other week.

After these two had been weaned, we awaited further Jersey heifers from Mr Morris, and in October we collected two more calves from him. Bramley and Russett were consuming Blossom's

milk and we had no serious scour problems with either of them – bottle fed first, then graduating to the bucket. Cello produced two weeks later, the Jersey×Murray Gray heifer destined for our freezer; and Cello was now to be used as a suckler until Blossom was dried off.

The market back at a reasonable level, Rick commissioned Don Gilbert to buy us a Friesian×Hereford heifer (Mr Morris had told us some tale that heifer calves take after the bull) when he was at Louth on Friday – the nearest market, twenty-two miles up in the wolds. That evening Rick collected a black and white calf, costing £66, from his premises. We put her alone in a strawed pen to settle down, offering only glucose water, and not introducing her to Cello until next morning. It took a few days before we no longer needed to watch over feeding operations, when Cello ceased to kick out at the small suckling creature; but she always showed preference to Sycamore, her own calf. Then the Hereford started to scour and although we immediately isolated her in a separate pen, before long both Cello and Sycamore were also scouring. With some haste, we called in the vet, a lady this time. Her policy was to inject with a long-acting antibiotic, to administer a drench twice daily and to feed with Ion-aids only for several days. Fortunately we had caught the infection early, so it cleared quickly. The vet advised us to inject future bought-in calves immediately with a 'Grovax' injection as a precaution against E. Coli infection. Cello was not an exceptional milker and the calf did not grow as well as a lone one might have; in future we restricted her to suckling one calf at a time. This calf fetched £96 back in Louth, six weeks later; not a great deal after deductions for veterinary expenses and petrol.

The two following calves, Hereford×Friesians, both bulls, each reaped a profit of £50 in five weeks. They both stood out amongst all the others in the market, and we got caught up in the excited atmosphere. The first was bought to go directly onto another suckler; it would be shown and got to slaughter weight soon after the year.

Blossom duly gave birth to Sunshine, a Jersey heifer, and as, at that time, we were very involved with marketing sweet williams, we again commissioned Don Gilbert to bring us two calves from the market, one for Blossom's surplus milk and one to put on Cello, intending to raise both to beef weight for sale. Being midsummer, the market prices were high and we had to pay £96 and £108 for two bulls that were only Friesians (which are no comparison to the Hereford × Friesian). In addition, one was already scouring from its stressful experience, standing hunched and dejected. On mentioning to Don that Cello and Russett were surplus to our requirements (our grass supply lacking during this dry spell) he looked them over and offered an acceptable sum, which now left us in the position of having an excess of calves for the available milk yield. Next morning Blossom was down with milk fever, and the vet was summoned in haste. After administering the calcium, she produced medication for Dibbit, as the poorly calf had been named. The scouring cleared immediately, but after a week, Buster (his boisterous partner) was afflicted, and for a short spell Sunshine too. It must have been at least two weeks before the calves were all drinking whole milk at the same time, but Sunshine lacked intelligence and was almost ready for weaning before she realised that a finger was not a necessary part of the operation. Powdered milk at £16 per 25 kg bag supplemented Blossom's surplus. Gordon, visiting on other business, slipped rubber rings on the bulls' stones, and the vet disbudded them.

On rich clover pasture, they were at last able to grow, with up to five months of grass still ahead of them this year. Then a big farmer in the village, stopping to buy vegetables, commented 'You'll need to keep them three years afore they'll be ready for the butcher. Them Friesians grow taller and taller in the leg and don't fill out to the end.' Imagining our money tied up for so long a period and the expense of getting them through two whole winters, we sadly put them back in the market. We had all allowed ourselves to become too fond of Buster and Dibbit. Luck was not with us that summer, the market had dropped and for all our trouble, they fetched what we paid for them (before deductions) and left us £52 out of pocket, but much wiser.

No more will we tamper with pure Friesians, we will beware the summer markets, and the little red and white Friesian × Hereford heifer we have now will not leave the holding until she is ready for slaughter, at around eighteen months; that way, no one else will take a cut of the profits, only us as the raiser and the butcher, who is both the slaughterman and retailer. We can feel better in the knowledge that this calf has not been moved on from place to place, that we have not been responsible for forming a link in that chain and hopefully that she has lived a happy existence with us and our other animals here at Fir Tree Cottage.

CHAPTER 18

I like sheep

Since we established the permanent pasture, we have been gradually building up our flock of sheep – we did keep a cade (orphan lamb) at the Marsh, but he did not do well. To feed hay would have made the meat very expensive indeed, and although we did not starve him, he never grew on the grassed area in the way lambs do here.

There are two methods of starting a sheep enterprise cheaply; one is to buy in broken-mouthed ewes, probably five to six years old – culls from a commercial flock, which can be bought for around £20 each. We have known people who have been lucky – taking lambs from them and selling them on in the market the following year for the price they paid for them, but things could go wrong. The sheep have, in the first place, been weeded out of the flock for a reason – mastitis, infertility, prolapse – and so these problems may recur.

The other way is to buy in orphan lambs, which is what we did. These lambs sell for about £7 at the time of writing, and we allot one tub of 'ewe milk substitute' per lamb (about two pints each daily) which brings the cost to £14. Of course, with a surplus of goat or cow's milk quite likely at this time of year, the cost can be considerably lower. With unlimited grazing once weaned, the lambs will grow to a good size for mating that first autumn (or for the freezer). There are, however, risks with this method too – from each of the three batches bought in, we have lost one. Two of the three cades we have lost, for two years in succession, have been tups we had left intact for the purpose of rearing our own male. This year we hope to exchange ram lamb for ram lamb with friends, to acquire unrelated stock. The third cade we lost just would not suck – I suspect that he had failed to receive colostrum in the vital hours after birth. He had already been rejected by three ewes when we took him on. We offered him to a ewe of our own, who had passed a decomposed foetus, rubbing his wool with her afterbirth and confining them together. He was certainly aware of her milk, but she just did not want to know. After three days had passed, we knew she would not accept him and carried him down to the warmth of the Rayburn. The inanimate form slowly revived, daily venturing further from his strawed box, but all was not well. He had been a sizeable lamb at birth, but now his ribs threatened to stick through his thin skin. His stuntcdncss was apparcnt against the rest of the jostling bunch, pushing and barging for their share of milk. He lingered on for three weeks, life fading from his emaciated body until finally I laid it with sorrow upon the compost heap.

When hand-rearing the cades, it is necessary to identify them in some way, so that, for example, we know which has been fed and which still needs feeding, so names are given in North American Indian fashion, solely as identification marks.

After weaning, the cades join the flock and easily revert to a wild, timid state. They learn to run from you instead of to you, and at this stage some form of holding pen must be rigged up before you do anything else. There are times during the year that the sheep need to be handled - lambing, shearing, dipping, drenching, and when checks are necessary. One of our first ewes to lamb had a stillbirth and had not cleansed. We, totally disorganised, had nowhere to corner or pen her in. Much time and energy was expended in rushing round and round an acre paddock. Since then, we have rigged up some movable hurdles to drive them into, but, until we have enough grassland and sheep to make a dog worth her keep (her feed would cancel out a couple of lambs a year), we will have to continue rushing around here, there and everywhere.

THE SHEPHERDING CALENDAR

TUPPING — This is where the year begins. The ewes are wormed and flushed (put onto the best pasture) three weeks before the ram is turned in with them, so that they are in peak condition for ovulation. Unlike the goats, we now serve the ewe lambs in their first autumn. When the original set were six months old, Tom Scrimshaw looked them over and declared: 'You don't want to put them in lamb this time around. Leave 'em a year and you'll have some fine ewes there.' So we followed his words of wisdom, gleaned from the experience of his maturing years. Next summer at the

shearing yard, Ron, his working years spent eking a living from his stock, remarked: 'Them ewes are over fat. You should have mated them at the back end. You'll have trouble getting them in lamb now.'

So we had kept those ewes unprofitably and to no avail for a year longer than necessary. Fortunately, Ron's predictions did not come true and they held to service at once. There was no noticeable difference between the first year lambers and those held back a year - both batches had singles, both had twins. The singles in both cases grew faster, just as one would expect.

For the past two years we have borrowed a ram, which means we get him when Ron has finished with his services, and this works well. A lot of farmers are now mating in order to lamb at Christmas or soon after, to make the utmost use of the subsidy received. We ourselves aim for March lambing, when spring is in sight and they grow along with the grass. Early lambing must eat into a fair proportion of the subsidy with the extra feed needed and the suitable delivery quarters which must be provided. In a severe winter, things could be very hard indeed. It is bad enough tending the adult stock with inches of snow on the ground, without the worry of newborn lambs at foot. We like to see ours frolicking in the sunshine. If the ram has been raddled, it is easy to see which ewes have been served by the 'red' mark on their bottom. By changing the colour after three weeks, those returning can be identified. A natural based raddle can be made up of a vegetable oil, plus a colourant, such as soot, and then smeared well into his chest and front legs.

IN-LAMB CARE — The ewes should be inoculated at six to two weeks prior to lambing with eight in one vaccine; and at twelve to sixteen weeks of pregnancy with Copper, against sway back. The gestation period is 147 days.

LAMBING — This must be conducted with common sense and careful observation. The ewe's udder must be checked for milk and any necessary juggling about done - a ewe with a single lamb given a triplet to adopt, a sick or weak newborn brought to the house. One also has to be watchful for signs of mastitis, and a ewe, standing around looking miserable days after delivery, could have a retained afterbirth. During their first week of life (it is illegal after this) the lambs have to be tail-docked and castrated painlessly with the elastrator.

SHEARING — The shearers accumulate at Ron Baker's farm in Stickford village and all the sheep owners in the surrounding neighbourhood bring their sheep to them throughout the day; the speed is incredible. We put our thirty pence per head into the kitty, thankful of the time it has saved us. These occasions always show that the community spirit lives on; everyone is so involved in his own particular task - the shearers must be kept supplied with sheep, there must be no hold-up in the conveyor belt system - that there is no time to stop and chat. With four or more ewes, one is under obligation to sell the fleece to the British Wool Marketing Board.

DIPPING — There are also laws governing the dipping of sheep;the compulsory dipping period for the eradication of scab was changed this year from September-November to July-September. Farmers had been up in arms about the late dipping, as in the cold weather and with heavier fleeces the sheep could easily become chilled. As with shearing, dipping is a communal effort; everyone contributes to the cost of the chemical. The sheep are thrown into the bath, fully submerged, then shake themselves in dog-like fashion and clamber up onto the slope, where the excess water runs off and trickles back into the dip. The Diseases of Animals Inspector has to be notified of intention to dip. On the arranged day and hour, he will turn up to inspect the dip, the chemical concentrate used and to supervise proceedings, stop-watch in hand, ensuring each individual sheep is submerged for the statutory minute. The old-timers do not hide their irritation with the bureaucracy, red tape and paperwork. Not only are we legally obliged to notify of our intention to dip, but in the week after the statutory dipping period, we must send in the form confirming we have dipped. All sheep passing through the market during this time, must carry dipping papers.

The chemical used provides protection against strike (blowfly), but it is in no way infallible. Surveillance must be kept for damp patches on the fleece, indicative of maggots at work. The tail area is a trouble spot; any dung collecting must be cut away and this is the reason that the tails are docked. We have known several new-

comers to this way of life, who have lost sheep through failure to dip – the maggots burrow through the skin and poison the bloodstream; the animal can be dead within a very short time. We have experienced the horrors of strike, so we will always dip in future.

* * *

The remainder of the year is less eventful for the shepherd; on the mixed smallholding, harvest is in full swing and thankfully it is just a case of keeping a watchful eye, moving the sheep regularly so they do not stay two Sundays in the same field, and avoiding the build-up of worms. Fat lambs leave the premises. Then autumn comes round and the blowflies are gone. The castrated ram (the teaser) is withdrawn from the flock, to slaughter for the house, the tup is brought in and the calendar has turned full cycle. We reckon on the lamb cutting out between fifty and sixty pounds at this stage. The meat is all frozen, the cats having the benefit of the head, pluck and feet this time and the fat goes for suet – it has too strong a flavour to be used as lard.

Our method used for curing goat and rabbit skins is no good here – the wool on the skin soaks up an excessive amount of formalin and thus it works out too costly by far. So, we scrape the skin as normal; then mix 7 oz alum and 3 oz salt together, forming it into a paste with the addition of a little vegetable oil, one or two eggs, $\frac{3}{4}$ lb flour and warm water. We rub this into the back of the skin and leave for four days. Then we wash, stretch and nail it to a board to dry, and sandpaper it.

Today our flock stands at fourteen ewes. We do not know yet if the grassland will support this amount (good land is capable of sustaining six ewes and their lambs per acre) in addition to what cows we have to turn out in the spring. I like sheep; with vigilance and careful management they take care of themselves; the offspring looked after by their own mother – none of the heartbreak due to their removal and the subsequent days of blaring that has to be endured with the milking stock. When we have our own ram, the set up will be complete; he making it his business to ensure each ewe is in lamb – how much simpler than the harassments of pig matings and movement licences, and even the AI service for the cattle. Yes, I like them very much.

VEGETABLE DYEING

When we visited the Alternative Technology Centre at Machynlleth, there were articles of the community's craft work on sale and a beautiful sleeveless Fair Isle jumper caught my eye, in subtle shades, reflecting the landscape around us. It was my introduction to vegetable dyes - the price of the jumper was prohibitive to our pocket (though cheap for the hours gone into it), so I vowed to begin dyeing my own wool as soon as we were home. I started collecting there and then; a bag of lichen from the damp trees - some of these are rare so do respect them. The bug had gripped me - now wherever we go, it is with a bag and the secateurs 'just in case'.

The amount of dyestuff needed does vary, but one pound to one pound of wool is the general average. If soaked overnight, the matter will give out the dye more readily and cut energy costs, when using electricity or some form of gas. The room is like a witch's coven when dyeing is under way; many containers with different brews and concoctions bubbling away on the Rayburn, all at different stages. Stainless steel and enamel pans are ideal; the latter can be picked up easily at a jumble sale or junk shop. Cast iron or copper pots act as mordants and affect the result. I will avoid excessive details, but the following will give the basic gist of the method.

WASHING — Once it has been skeined, the wool must be washed to remove the grease. Insufficient washing will produce a patchy effect.

To avoid felting, the wool should be gradually accustomed to temperature change - not plunged directly from cold to very hot water and vice versa. The mordant or dye should be maintained at just below boiling point at all stages; and the wool should only be stirred once or twice during processing. Soft water is preferable.

MORDANTING — In most instances, the wool needs to be treated with one of a variety of chemicals to fix the dye. These chemicals are known as *mordants* (beware - some are poisonous). The wool can either be mordanted prior to dyeing or actually during the dyeing process. The wool should be exposed to the mordant for forty-five minutes to an hour.

The mordant I now use most often is alum. Use 3 oz alum, 1 oz cream of tartar to 1 lb wool. The cream of tartar is added to give the wool lustre. Alum gives the true colour of the dyestuff and is also the cheapest.

The following chemicals can be used as mordants or treated as additives, as they alter the colour in some way. Used alone they are a dye in themselves. Tin (which brightens) and iron (which darkens) should be used as additives only. Copper sulphate (2 oz to 1 lb wool as a mordant) and chrome (½ oz to 1 lb wool as a mordant) can be used as additives or mordants. Chrome is said to emit toxic fumes, so I work in the open with this chemical on a camping stove. It is also light sensitive.

If an additive is used the wool is previously mordanted with alum, then a small amount of the chemical (between 0.2 oz and 0.4 oz to 1 lb wool) is added to the dyebath to alter the tone, that is to darken or brighten. It is preferable to use iron and tin only as additives, as larger quantities tend to harshen the fibres.

THE DYEING PROCESS — The dyestuff can either be simmered beforehand to extract the dye, or the wool can be added simultaneously. Barks and other coarse matter that need longer treatment should be simmered alone at least for part of the time. Dyeing takes between half an hour to two hours maximum - prolonged exposure can have an adverse effect on the wool, it can even disintegrate. A muslin bag to hold the matter saves the endless job of picking off the 'bits' afterwards, such as dandelion petals or gorse.

RINSING OUT — Now the fun! The nice thing about dyeing is one never knows what colour to expect. I have included a table of plants I have used to date, and the colour they gave me (*see* Appendix). Rinse the wool thoroughly, first in water nearly as hot as the dyebath, gradually decreasing the temperature until the water remains clear, usually after about four changes. Do not use running water though. Then, I hang my skeins on the line in the yard, the shades blending naturally with the surroundings.

I have found that each season has its own particular shades:

Winter Buffs and full colours.
Spring Pale yellows and greens.
Summer Brighter greens and the colours of the flowers. Some quite dramatic colours can be obtained from garden flowers, such as dahlias and chrysanthemums.
Autumn Pinks and purples from the berries. These are not very fast, although the addition of vinegar is said to improve their stability.

The seasons cannot be interchanged. It is as incongruous to take 'lichen yellow' from my winter calico bag and mix it with sweet william green from my summer bag, as to see an apple tree in blossom in a field of corn stooks.

The amount of dye held in a plant is related to the hours of sunshine absorbed. Hence, far more vivid colours can be expected from plants gathered in warmer climates, the same colours as the exotic flowers on picture postcards. It is hard to obtain a true blue in this country - except from woad and indigo which require a special fermentation process.

Many variations can be added to the basic dyeing process described above. These can include tie-dying in the skein; top-dyeing (placing for a short time in a second dyebath of a different colour); dyeing white and dark wool, for example, Jacob. Both colours are carded together prior to spinning, then dyed. Jacob fleece dyed in cutch produces purple. A bright colour that has been dyed in the fleece can be toned down by carding with white fleece; likewise, a tweed effect can be achieved by carding together two fleeces, dyed in different colours. I knitted a multi-coloured striped cardigan by plying lots of variations of two different coloured skeins of wool together.

The dyeing bug led me on to growing plants especially for dyeing. I did buy in various dried plant materials from a commercial source (bloodroot, saunderswood, safflower, logwood, fustic) but somehow this seemed to be cheating, and as I already knew the colour to expect, a little of the excitement and satisfaction was lost. Cutch incidentally gave excellent value for money - without using a mordant, 2oz dyed 24oz wool. The colours ran through browns to pinks, and even after six dyebaths the colour was still going strong.

We had been spinning our own wool for several years on the Ashford wheel, but only using white fleece and sometimes Jacob. Now a whole new dimension was opened up to me; I love experimenting with colour in knitting. We are not perfectionists, our wool definitely still has that homespun quality, sometimes lumpy, then verging on dangerously thin. From here I went on to using my own designs - maybe too elaborate a word for the simple garments that are turned out from my size one or two knitting needles!

It must be remembered, that it is impossible when using natural dyestuffs to ever reproduce exactly the same shade again, so to make a garment in one colour alone, a large enough container to immerse all the wool at one time is needed.

We have also played around with various weaving looms - a small loom using bended willow boughs, and an enormous contraption Rick put together in our bedroom, based on the Navajo design, which left little space for us to sleep and gradually collapsed as Rick used it. We now have a twenty-four inch table loom. We practised first with straightforward mats, bags, curtains and wall hangings, where the scope for experiment with colour is endless, with different warps and also different wefts. Now I have ideas crowding into my head for woven garments, as winter evenings draw in again.

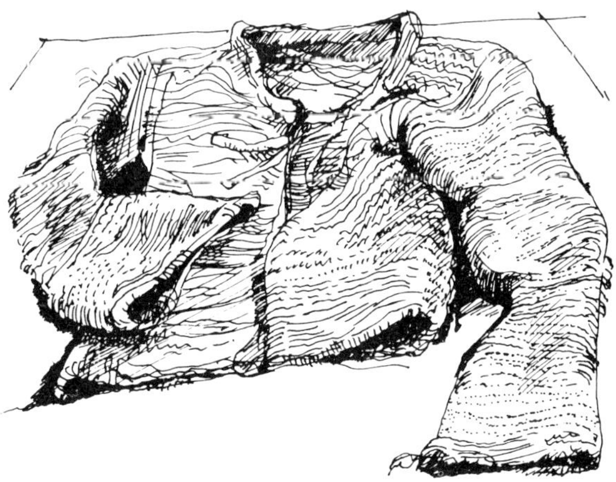

CHAPTER 19

Down the dyke system

Over the years, we have tried rearing a variety of poultry. We used to buy in goslings for a couple of pounds each at Wisbech, to raise for Christmas. We ran them on the lawn, underneath the gooseberry bushes - anywhere we could find a patch of grass - until we finally established the paddock; and except for the final three or four weeks, when we fed them on barley meal, they cost us nothing. Not that they got up to any of those fantastic weights, 30 lb or more, you hear of.

The first year we were not successful; we ran on four, intending one for our Christmas dinner, and three for sale. It was 1976, Dicken had just been born and I had been home only a matter of days, when the big Christmas poultry market at King's Lynn became imminent, so Rick had to pluck the birds.

It was in the days before we heard of wet plucking - now all our poultry are dealt with by plunging them feet first into the washing machine full of water just below boiling point and held there for about a minute. The feathers will then literally fall out into your hands - but beware, if the water is at too high a temperature, it can cause splitting of the skin. After Rick had laboriously removed every one of the outer feathers of the geese, he then discovered he had to go over it all again, plucking out the undercoat of light down. It took, without exaggeration, hours for each one. It was long after dark when Rick came in, chilled to the bone, with three stiff, featherless geese hung by their necks - and then they still had to be drawn.

The next day, at 8.30 a.m., Rick delivered the geese to King's Lynn auction, putting a reasonable reserve of fifty pence a pound on the birds. At 10.15 a.m. he arrived home to yesterday's chores; the market had been hectically busy, hence the delay. After another very late lunch, Rick telephoned the market to enquire what price the birds had made and was informed the reserve had not been reached. So he had to go back to King's Lynn to collect them. The geese had only been fetching £2.50 each (the going price for goslings was £2 at two weeks; not much to cover the hassle of dressing them) when everyone else we knew was charging, and getting £1 per pound.

The next year was even worse. This time we purchased eight goslings, in spite of our previous experience. Until they were fully feathered, we housed them in the goatshed at night and carted them to the gooseberry patch, provided it was reasonably fine, by day. One was had by a rat as it slumbered in the milking bay, huddling with its companions; another remained stunted and gave up the fight; a third was had by a dog in its pen during the day. The other five grew on, developing into magnificent birds, appearing proud and arrogant, as they flapped their wings and pulled their long elegant necks back. They were now hardy enough to live permanently out of doors with access to a straw-bale shelter. Their

hissing and honking should have been warning to anyone within half a mile radius of approaching danger, living up to their guard dog reputation. But one morning I went out and there were five little corpses - a fox had been. The bodies were barely mutilated; one had just been dragged a few yards and left. It was clearly a case of killing for sport rather than necessity. On still nights the eerie call of a dog fox could often be heard, but it was the only visit he made.

It was a case of third time lucky. From then on we kept a permanent set of two geese and a gander (geese mate for life and are good for twenty years or more). One of the geese was a Chinese, noted for their prolific egg production, and the others Embdens who throw a large carcass in any resulting offspring - which to date have been few and far between. We do not seem to have a high fertility rate amongst the eggs; perhaps the provision of a pond would do something to alleviate the problem. Geese prefer to mate in water.

We did incubate some eggs once, religiously splashing them with water, imitating the mother goose waddling back from the stream; but without much success. Now we leave the job to the geese. If they go broody and hatch some, all well and good. We do not need too many; after all, six geese eat the equivalent to one sheep, at which rate their meat is fairly expensive and also has a high ratio of bone to flesh; but we do aim for as much variation in our diet as possible.

So far, the geese have not proved good sitters. As soon as the Embden showed signs of broodiness, the other joined her, but half-heartedly. The Embden would sidle up and pinch a few eggs, but then she would wander off and be gone perhaps a day or more, so the embryos in those eggs would die. The Chinese sat tight, except for the harassment from her neighbour; every now and then, it would get too much for her and she would shift along with only some of her eggs and they too would be spoiled. Of fifteen eggs, not a single offspring resulted.

We generally just have the eggs for the house (although we once sold some as believed to be fertile at twenty pence each), but geese lay March to June on alternate days, so the eggs come at the time of the spring flood from the chickens!

The only time I would not trust the gander is in the breeding season, when the females are either laying or sitting and he

becomes fervently protective towards them. Once he pounced on Clyde and knocked him flying. Before I could move, the gander was on his back, pecking his neck. Not surprisingly, Clyde is absolutely terrified of them now.

* * *

Then there were the ducks. At the Marsh, they fitted well into the pattern of things. From necessity, all the poultry were penned and, on proprietary feed, the ducks proved to lay with far more consistency than the chickens. We had chosen Khaki Campbells for this reputation alone, but although the carcass was in no way comparable to that of the Aylesbury, the surplus male birds that appeared on the dinner table were certainly deliciously edible. The polecat finally put an end to the good egg strain we had, and a long duckless spell passed.

Then, one bleak winter's afternoon, we looked out to see an unfamiliar visitor on the lawn – a Muscovy duck – and since that day, he has stuck around. Little by little, he became bolder, until at the clunk of a bucket, he would arrive with a flapping of wings to gobble grain alongside the hens. Winter turned to spring and it became apparent that 'he' was not a he at all – by the discovery of a clutch of around twenty eggs in a feather-lined nest, in a disused rabbit hutch. Twice yearly since then, we have been presented consistently with two clutches in payment for her keep. Not being fertile, we take the eggs away for the house, but that first season, we replaced them with china eggs and encouraged her to go broody. Then we bought in a dozen very expensive day-old Khaki Campbell ducklings from a reputable source and gradually, one by one, introduced them to her. She was completely free to come and go as she pleased, and this she did, trailing her brood with her. She was not a particularly good mother and soon the numbers dwindled, as presumably one became left behind over here, another lost over there. When she had thus depleted the flock to just five, she took them down the dyke system and did not come back.

The book informs us that provided geese and ducks are fed at regular intervals, they are happy to fly off to water by day and return like homing pigeons at dusk. All summer they travelled around the fen on the waterways. Those five ducklings were not only all that remained of our £14 initial outlay, but they had also eaten through quite a lot more money in feedstuffs, ducks being greedy creatures.

When autumn came, there was the danger that mother duck would abandon her charges, who would become adult ducks, never knowing they were really domestic birds and their home was with us here. We donned plenty of clothing (there was a definite nip in the air) and drove round, along the rough farm track, fields ploughed or showing signs of sprouting wheat on either side. Rick and Jade pulled off shoes and socks in readiness to wade in if necessary. We threw clods of dirt, stones and made noises in order to drive the ducks up the bank, where they were immediately pounced upon. It was quite a task and when we got them back, we confined them to an outdoor pigsty - where the mother duck still flew in and out as she pleased, but the walls were prohibitively high for the youngsters. As soon as the drakes were of suitable size, they were dressed for the oven, and one went for a raffle prize. There were just the two females remaining. We realised that future eggs from these ducks, that had to be permanently housed, would be costlier by far than the eggs from our free-range hens, and so, when the opportunity arose, we bartered them in part exchange for a table loom. That saw the end to ducks here for good, as they do not fit into our system.

* * *

When we first bought turkeys, it was with some apprehension, as they have the reputation of being frail creatures, teetering permanently on the point of death. To lessen the risks, we purchased poults at six weeks rather than day-olds. From then on, the method of rearing that we have arrived at by trial and error is confinement for a short spell on proprietary crumbs, with anti-coccidiosis additive, until the danger period has passed, followed by free-range foraging for the length of their growing lives (with just a handful of meal to bring them in at night). It is a delight to watch the gentle creatures peck-pecking at the seed pods and the vegetation above ground, not grovelling in the dirt like the hens, and, when approached, they will stay to communicate, not fleeing in a flutter of feathers with an alarmed clucking. The last month of their lives, when the weather is not so kind and there is not much to glean from outside anyway, they are happy to stay in the welcoming warmth of their strawed-down shed (under lock and key, as there was quite a spate of poultry thieving in this area the Christmas before last) and to be fed as much barley meal as we can possibly give them.

The poults hatch on about the longest day, and are plucked on about the shortest day. The cats waddle round with bloated bellies for days. It is, for us, a nice part of the festive season; folk dropping by for their orders, tempted into the warmth for a glass of wine and a mince pie. Everyone is full of good cheer, and it puts us well into the Christmas spirit. Most of the customers ask for a small or medium bird, so we have stuck to hens - the 25-40 lb possible with stags is just too much for most people.

Our initial attempt at turkey raising went badly, not because of their feeble constitutions, but rather the fault of outside forces beyond our control. One was taken from out of the shed by a two-legged fox (when little more than a large chicken). One suffered a broken neck by venturing too close to a cow's clumsy hoof, but we salvaged a ten pound carcass. The third casualty was one that remained smaller than the rest of the bunch; it took on a typical dejected appearance - hunched up, head down, frequently hanging back from the others. We had read dire warnings of what happens when turkeys have access to chicken terrain and suspecting this was the dreaded blackhead, picked up from the worms of hens, culled it, cutting our losses there and then rather than pushing further food into it and having it die on us. Our fears were confirmed by the mottled liver, though we enjoyed the meat; but afterwards we heard that the disease is not only preventable by medication in the water, but curable by means of a larger dose.

The remaining three of that first bunch of six reached good weights, 17, 18 and 18½ lb, and even taking the losses into account, we were able to sell the meat at a reasonable profit. They came into lay a week before the slaughter date, with their white speckled offerings, but we had heard how difficult it is finding a suitable stag (the stags produced nowadays are far too large) which is why AI is used amongst commercial breeders. So we had to be hard-hearted. Our resulting oven-poultry, fattened on barley meal, certainly needed no injections of fat to keep the bird moist for basting. In fact, wrapped in foil and cooked at a low enough temperature in the Rayburn, they needed no added fat whatsoever.

This year we doubled the venture - a dozen poults - which

proved so popular, we had to keep turning custom away. Nothing went amiss; they averaged 14-15lb. In fattening fifty cockerels and twelve turkeys, we had made two turkeys and eleven cockerels for ourselves, along with a couple of pounds too, so we will raise the numbers again next year - not by too much though, because the feed bills become a nightmare towards the end.

The cockerels mentioned were Ross Is, bought as day-olds and given to broodies; one eventually putting claim to all fifty as usual! We had our own Rhode chicks of comparable age - supposed to be a heavy breed themselves - running with them, and they did look small - perhaps a third of the size of the muscular white birds with their massive legs in readiness to support the fifteen pound frames when fully grown.

Two of the home-produced chicks happened to be particularly dark and one morning we went in to find the other cockerels had shown their colour prejudice at a very early age and turned vultures overnight. The poor victims had been rendered featherless, exposing bare flesh for the whole length of their backs. They must have been in dreadful pain and we expected them to die, but in solitary confinement, with fresh air and tender care, they slowly recovered.

The birds had been recommended to us, as not prone to heart failure to the same extent as other of the large breeds. In fact, we did not lose any from ill health whatsoever; just one did not end its days in the way we had intended - it fluttered into a pen of fattening pigs with the intention of raiding their food trough, and could not flutter out quickly enough before they realised intentions of their own. But the cockerels began to eat us out of house and home, and eventually we managed to sell a dozen, which gave us a bit of extra capital to fatten the others. Then we began selling some of the largest off as dressed birds, when they reached about nine pounds live weight. The butcher came and took a large bunch which we did not even have to pluck and that left us with some Christmas orders to fill and the remaining eleven of suitable size for our freezer as profit.

CHAPTER 20
Natural disasters

One can prepare the soil, with plough and harrows to the right tilth; sow the seeds at the correct depth and at the correct distance apart; tend those seeds with care, fertilising or manuring as one sees fit and keeping each plant free from any weeds; but having done all we should do, we are not assured of a bountiful harvest. Our crops are entrusted to a far greater force - nature. This power can destroy all our efforts overnight - by gale force winds, excessive rain, too much sun or late frosts. Not until all the harvest is safely gathered, can we celebrate. The fens with their bleak openness, hold the scenery in the everchanging sky. Without distant rising hills and few trees, there is nothing to intervene or interrupt it. This whole force of nature can be felt in the vastness of the sky.

Each year, different crops succeed and fail. Hay is particularly susceptible to Mother Nature's hand; the food value it contains depends on the stage at which the weather permits it to be cut; it is left in her care for a week, at which time excessive rainfall can render it useless.

The farmer's lot is a hard one; besides the fate to which he surrenders his living, there are other adversities to contend with:

> One for the pigeon, one for the crow,
> One to rot and one to grow.

Such mishaps, for example, as when all the cades escaped and trimmed up the thousand young runner bean plants intended for the 'pick your own' harvest. Over the years that we have been smallholding, I think we must have now experienced a taste of all nature can hand out - drought, gale, blizzard, severe frost, excessive rain.

We came to Ivy Cottage when the long drought, which lasted for two summers, was in its early stages. We had not really been through a summer of normal growing conditions to have anything to compare with, and when the long, dry summer went into a mild winter with little rain, we thought every winter was like this. Every day throughout those months, we worked digging couch grass out of an area cleared of gooseberries that had been left too long uncontrolled and unpruned. Any slight frost that had made the soil clod together rapidly cleared, and, arming ourselves with forks soon after breakfast, we would work until the light faded and was gone, at around 4.30p.m. Only then was the fire lit, and all we burnt that first year was broken Holland trays, tidied up from around the holding. We did our bit to conserve water and watched the sweet corn grow. Unfortunately we had not yet invested in bees. The strawberries suffered the next year - lack of rain meant lack of top; the fruit ripened small, red spider was rife and the season short.

It did not affect us unduly however, being on mains water and without rationing. We put our brick in the toilet cistern and although we did not possess a bath anyway, a stand-up wash at the sink was no hardship in such mild weather. It was not until we read *The Great Drought 1975-76* that the full meaning of the eighteen months without rain was brought home to me. The drought ended in September, the only autumn we quit fruit picking early, due to the unceasing miserable rain.

Looking back, the next two freaks of the weather occurred at the turn of the year, when little was in the ground to be spoilt. On 2 January 1978, with Jade somewhere between Peterborough and Walpole returning on the school coach from a pantomime, a tremendous storm blew up. It wailed and roared with violent strength, gushing down the chimney and rattling the windows alarmingly. All night the gale's destruction resounded in the Marsh - the shattering of glass, crashings, bangs and clatters as objects were battered, flattened and swirled around.

The next morning, the storm having subsided, we surveyed and cleared up the devastation. Many homes in the vicinity, often relatively newly built, had lost chimneys; although Ivy Cottage, 200 years old, was unscathed. It was not until we turned on the radio news that we realised how hard the storm had hit the area, for the River Nene had burst its banks at Wisbech and many homes were evacuated.

I have already told how the blizzard of 15-16 February 1979 affected us personally, and tucked away in a hospital bed, I was not directly involved in the clean-up operations; but it is not until times of crises like this, that the true community spirit shows itself.

All able bodies helped with snow clearing, whether by shovel or more mechanical means, and Greg got his lorry out and made it through the snow on the second day, after several tries, to fetch milk for the villagers where the dairy had failed. The large proportion of pensioners were cared for, their paths cleared and a check kept on their provisions.

Two years later at Stickford, it was April, Bryony was three weeks old and all our crops were sown, when we had three days of continuous driving rain, from the north-east. All day Friday it rained, through all of Saturday and as I fed Bryony during the night, the rain still lashed unceasingly against the windows. Suddenly, I heard a strange sizzling sound coming from somewhere below and then everywhere was black. Rick, shaken awake, felt his way downstairs to the candles and investigated. An ugly black area revealed that the meter box had suffered fire damage and the trip-switch had blown. The electricity board assured us all was safe and we joined their growing list of emergency calls for the next day. I snuggled baby down into our bed covers – the bedroom storage heater would operate no more tonight.

The scene from the landing window was unbelievable when daylight crept in. Two-thirds of our little world was under water and still the rain fell. We could do nothing constructive outside until the weather had cleared up, so we concentrated on making a cup of tea and breakfast on the camping stove. Then we went out to attend to the animals; Blossom was snug and dry in her winter quarters again so soon after being turned out to grass; we dashed back across the slushy yard and indoors to sit around the Rayburn's warmth, sewing or with book in hand, waiting for the electricity board to arrive. Just before lunch, two large vans pulled up and an army of men, equipped with helmets, ladders and tools, filled the pantry where the meter box is installed. The cause of the fire was found to be driving rain, forced through the landing window and dripping onto the meter below, which made the trip-switch blow. The wiring was repaired during the afternoon, by bypassing the trip, and a new switch was installed later that week.

The rain eventually abated and all farm-hands were provided with a spade, shiny new or recovered from the dusty depths of some farm building. The task was to get lying water off the crops, so that the least possible damage occurred. We followed their example, digging courses for the water to trickle down to the overfull dykes.

Our field to the west of the house was the main problem area – the mangolds and fodder beet were on the bed of a lake and the potato ridges were streams. Large puddles lay in the strawberries under the winter wheat and the onion block on the vegetable garden was completely submerged too. It took days for the water level in the dykes to subside sufficiently to drain our land.

Surprisingly, a lot of our vegetables did grow and many did well. We had onions, although a reduced yield, and the fodder crops were good; only the potatoes failed completely; of seven cwt of seed planted, only a half cwt was harvested – the error was ours in not sparing the cash to replant. Because of the heavy rainfall, the foliage on the strawberries (first year pull) grew and grew – then the strawberry season being hot and dry, a bumper crop followed, yielding 6,300 lb from the half acre. Had rain fallen on and off too during this hot spell, the situation may have been vastly different, as in the succeeding year – so fickle is the hand of nature.

The hardest winter we have endured was that of 1981-82. There were two prolonged spells of bitter cold, the first with snow, the other with severe frost, when temperatures reached record lows. Water for the stock had to be carted from the bathroom tap, slopping a trail in its wake. The cows chomped through their dry diet of hay, barley straw and concentrate, as the kale was standing stiff in the field and the roots were a solid mass even in the clamp. The free-range hens crouched miserably, combs bent and bluish, awaiting the rattling of the corn scoop; their only pleasure now being a scratch around in the straw barn or beneath the tumbrils in the cowshed.

Our energies were concentrated within the house, even the muck-spreading had to be abandoned, so we attacked the decorating for the first time. It was almost two years since we moved to Fir Tree Cottage, and to date all spare capital had been invested outside. Loft insulation, the installation of the Rayburn and the conversion of the two small sitting-rooms into one large, more practical living-room, were the only improvements until then. We had just about completed the kitchen and toilet before the money was gone and the thaw was upon us, when we had had enough of being confined to the house anyway.

Most of the time during that hard winter, we hibernated, and really did not know anything but our own business, isolated down Scarborough Bank. Rick helped out at the pig farm during the

mornings, although now officially laid off, while the full-time worker was away on holiday. Once our animals had been attended to and the tea things cleared away, we would settle down to long evenings, doing those things we liked best. It was an effort to go out. We did get through ten miles of appalling road conditions to the winter solstice celebrations, which that year comprised a meal in a village hall. A grand raffle was held at which everyone present donated a prize - a live Khaki-Campbell drake was our contribution. All the children wore fancy dress; ours became clowns with Bryony in cardboard wings as a butterfly. Sandra and Rob invited us down for Boxing Day tea, and we drove over to Dad's, in Suffolk, over Christmas, but mostly we watched television - a rare treat - on a borrowed set, and played with the children's new games.

The cold months passed and spring returned with all the activities it heralds, and as the hedgerows changed to green and the dykesides grew up and the rogue weeds made a haze upon the once bare earth, so too did those seeds spring up that we had so tenderly sown. The sun was warm and soon everywhere looked green; the roots of the bought-in tomato plants had obviously found the chicken manure and the kale stood sturdy and thick. I was busy chopping it out with the long hoe into bunches eighteen inches apart. Rick was endlessly earthing-up the potatoes and the next day yet more would be bursting out of the soil. But such bounty as the crops promised was not to be - one night, Jack Frost intervened.

As soon as I discovered the severely sharp frost the next morning I sprayed the tomato plants with water, but when the midday sun had beaten down, their leaves looked wilted and black, the potatoes were the same - all trace of growth gone. The kale seemed ruined and the hours spent hoeing, wasted (although after later rains, seeds lying still dormant, germinated, to produce a fair bed). Even the weeds were checked; the dock and sorrel leaves already high and vigorous, were sad and singed; but it was the soft-fruit growers that had suffered most. There is no insurance cover against frost, it being impossible to ascertain the extent of any damage. It was some time before we could see the effects on our soft fruit. The raspberry canes were brown and dead-looking above the mass of new growth of next year's canes. The early strawberry flowers failed to develop into fruit - and this was only the beginning of our troubles.

May and into June continued hot and dry. The grass did not

grow and the animals were always on it, so much so that Rick began cutting roadside grass for the cattle. Then the weather broke and for weeks it was humid with intermittent showers and often long downpours too. The weeds raced ahead of anything else and the strawberries grew terrific top-growth. One day as we continued with the seemingly endless task of weeding the main strawberry block (coming up for second year pull), we noticed the tell-tale furry signs of botrytis, contagious rot. It seemed very widespread. We consulted the Ministry, as the crop was unsprayed. The advice was that a fungicide should have been used in a comprehensive spraying programme of three applications as a preventative measure only, prior to fruiting. But we were organic. The unusual amount of top-growth prevented the air from circulating around the trusses, made worse by the row widths. The strawberries had produced many runners which, although checked by sloughing the alleyways in the winter, now appeared as a continuous field without rows. Moreover, the poultry manure mixed with sawdust so lovingly applied, now provided a permanent moist mulch upon which the fruit rested and so rotted. Later we were to learn this nitrogen was our downfall, in producing the excess of leaf. The botrytis attacked early and so we watched whole trusses rot and fall off, rather than just the strawberries themselves. It was not until we entered a completely dry spell, that we pulled any fruit at all and then we only salvaged two per cent of our previous year's total, whereas the Ministry quoted us a sixty per cent loss of crop amongst the commercial growers using chemicals.

Others were hit too. The humid conditions were ripe for blight and we heard of farmers lifting main crop potatoes in July where the crop lay wet, to cut their losses before rot set in.

So it came about that this marked the end of our high organic ideals and a depressing period followed with much thought and soul searching until we got to our present situation and arrived at a compromise. I believe that we do not shape our own destiny; but can only do the best with what we are given.

CHAPTER 21
Grass is a crop

It would have been a waste of our precious land on the Marsh to grow grass, and stock was forbidden on the allotments. We did sow a small patch where a derelict barn had stood and where the soil was of no depth, but this was solely as an exercise area for the goats. Later we established a ley by undersowing wheat in the first pen that Grace, the sow, had cleared; the geese lived here. But in the main, our animal feedstuffs were imported and, each evening throughout the summer months, Rick would take his scythe in the car down to the fifty acre open sewage farm at the back of us, and fill it with long, lush mown grass to feed fresh to the goats after milking (a sure way to increase their yield) and for the rabbits. We were further able to increase our acreage by browsing the goats along the wide roadsides.

We were not then living in a stock area. Although there was Len Powers, who kept a few bullocks on his hundred acre mixed farm (grain, potatoes, apples, strawberries, a few geese and even some mangolds). It was from Len we bought our bedding straw and generally hay, although Rick had two attempts at his own. The first came about after the discovery that the river road had been flailed. I do not know what month it was, but it was a very wet one. Rick disappeared each day, after heavy rain had fallen during the hours of darkness, to turn the 'hay'. This went on for some time, until even Rick could no longer claim that the vile, black, decaying mass had any resemblance to hay. The next try – another flailed roadside – met with more success, in that at least Len baled it; the amount was insignificant, but meant a lot to us with our goats in those days.

At Stickford, with much more land to play with, we immediately put two acres down to grass – tethering Blossom and the goats on the dykesides while it grew. Then the additional two acres set aside for barley and wheat were gradually broadcast with a permanent pasture mix too, in the belief that the money made from the sale of one lamb would cover the price of bread wheat for the house. On our good land, six sheep and their lambs can be stocked on one acre, whereas the yield of wheat from our method of growing and primitive means of threshing was absolutely minimal.

Grass is a crop and must be treated as such. It is broadcast in the autumn or spring, either by undersowing or by itself. Undersowing is the scattering of the grass seed at the same time as the grain. Then, when the corn is harvested, there is a thick growth of pasture beneath all ready for grazing and thus time is saved.

Once in, the organic husbandman must watch over the crop to ensure it gets ahead of the rogue weeds. Willow weed is our problem here, a weed of arable land, and whenever we put in new pasture (which we do with a fertiliser spreader, which is far easier and gives a more even distribution than the fiddle drill in our inexperienced hands) this weed takes over and has to be mown off to allow the grass to grow up, smothering any further alien seedlings. Similarly, a poultry ark stood in one place too long left a bare oblong in its wake, and, before the grass beneath the chicken manure had a chance to break through, the willow weed had re-established itself. However, once the reaper had been through, the grass took over again. Any thistles that spring up, the cows will ignore, but when mown and wilted, they tuck in with enjoyment.

We aim to sow at the beginning of April, when the land is ready to take the tractor, and there is still likely to be plenty of moisture to give the seeds a good start. Under fair weather conditions, spring sown pasture can be lightly grazed with the sheep in six weeks; this will have the effect of tillering - pushing out more shoots. Rolling does likewise. Autumn sown pasture is not touched until the spring, but is all the better for its earlier start. Established sward can do with harrowing once in a while, to rake out the dead undergrowth; treat grass roughly and it will spring back for more.

The ideal situation is the farm of mixed stock, where the worms can never build up, because the animals are rotated one after the other around the small paddocks - areas only large enough to hold the stock a few days, before moving them on. Hence the saying 'Sheep should not hear the church bells ring twice in the same field.' Thus, calves (eating the longish grass) should be followed by cows then by sheep (who nibble around the cow pats and close graze), followed by hens (who scratch and distribute the pats).

We have found the hardest part of grass farming to be maintaining the correct balance between under- and overgrazing, the

latter being at least as harmful. In fact, one of the general conditions of 'summer keep' – rented grazing land for the summer months – is that the tenant is held responsible for this. The problems to be taken into account are that the animals are increasing in size and appetite throughout the season; an April born calf will be consuming quite a quantity of grass by August, having replaced milk as the main source of sustenance, thus the month in which any calves are due has to be born in mind. Together with this, but completely out of our hands, is the interference of the weather. A period of hot days without rain means the grass does not grow and the animals soon get on top of it. The sheep constantly on their feet is a sure sign that the pasture is being overgrazed. When the hay bales have been led in, that field can be reintroduced, which helps until the September flush comes and our problems cease.

I should mention something about fencing here, but I think the least I say about our venture in this area the better! The trouble is, money has always been scarce – all we had has gone into buying Fir Tree Cottage. Both time and money are usually in short supply. Posts and fencing are now very expensive commodities and we have tried to save money on them, by bringing our used fencing from Ivy Cottage, buying second-hand at farm sales, scrounging bits and pieces here and there, and so on. This, together with the fact that when a hole appears, there never seems the time for thorough repairs means that much time is spent chasing escaped lambs, leading to frayed tempers and arguments! One thing we have found, though, is that when newly fencing an area, the only really satisfactory way is with a wire strainer. Goats are an impossibility to fence in. On coming to Stickford, we free-ranged them initially on our two areas of new pasture, but, before long, the vegetables growing on the other side of the fence looked far greener than the rich succulent clover ley and we resorted to tethers once more.

All went well in 1980; the sheep stayed confined in their allotted area and Blossom was the only cow we had – slow and sedate even as a heifer. When we introduced Cello to the herd, together with calves born or bought in, their fun and games caused us to tack up scrounged or second-hand barbed wire too! At present, we have four one-acre fields, which would be better managed as half acre paddocks, but money restricts us. Ideally, we should invest in an electric fence unit to strip graze, and in this way, any fraction of the acre can be set aside for cutting; and if a dry spell leaves the stock short, for example, yards of the hay field can be reclaimed. Incidentally, Rick has discarded the post-hole borer; he now finds he achieves a better fit by just boring the hole with a metal stake and sledge-hammering the post into position.

We estimate our winter stock requirements of barley straw and hay on the generalisation that the cows are in six months of the year and out six months, the weather being the deciding factor. The first mild days of late March and early April, we begin turning them out in the day - watching out for signs of bloat and taking precautions against magnesium deficiency. The grass is at its best early spring and early autumn (the September flush) which boosts the milk yield; the wrong months to have a cow dry! From about 20 October, we house the cows at night - milking is more pleasant in the cosy shed when the clocks are put back and darkness comes early. Anyway, left outside, food would go into keeping them warm, rather than put into their milk. Pleasant days still see the cattle out grazing in November, but after this the pasture becomes poached, patches trodden into mud, and the goodness is gone out of the grass.

The cow eats the grass; the by-product she spreads as manure while she grazes, but the remainder goes into the milk, which feeds our family and the calf to be exported from the holding. To replace this outflow, extra fertility must be brought onto the pasture. We do this by buying in bedding straw, roadside hay and limited concentrates, for the half year that the cattle are indoors. Having rotted down, it is carted by tractor, trailer, fork and muscle to be spread over the grassland in the winter months until one affluent day we pick up a muck spreader at a farm sale.

The sheep stay in the field all the year round. They take the growth right back to the ground and improve the pasture. In snowy weather when the grass is hidden, these animals will scrape to unearth green tufts and nibble that area bare. In a particularly hard period, we barrowed hay and concentrate up to the top field, but it was several days before they realised this was edible. We have read stories of Welsh mountain sheep actually starving to death with the unfamiliar concentrates before them.

Hay should not be taken from newly-sown pasture, so it was with great enthusiasm that we awaited the 1981 crop. Each day we

consulted the Meteorological Office's long range weather forecast for a settled spell. One fine Sunday, Rick began work on the acre set aside with an Allen scythe, while I followed behind, pitchforking the fallen grass into winnows. Because the grass was lying, instead of taking a breed around the circumference, it could only be cut from one side. This meant I had to race along to finish at the same time as the machine, as it roared noisily back and forth, or else operations would be held up on the return cut. Also, if a full cut (fifty inches) was taken, the grass clogged on the cutter blade, therefore Rick was only making a half cut and so progress was slow. The field left was not close-cropped, and George Sharp later blamed our setting-up of the blade. That night, the radio forecast rain, so we made the grass from the third of an acre already done into tripods. By the following evening, the remainder of the field lay in neat winrows and then our nightmare began.

Being totally ignorant on such matters and going by the book, each day with the surface dry to touch, we turned that hay by hand. Each turning took several hours of hard work. Where the romantic notion associated with haymaking comes from, I do not know. The sun's heat was unbearable and I had a choice of wellingtons, a new pair of handmade sandals or bare feet. I chose the latter and consequently every few feet, trod on a thistle, wilting inconspicuously. Blisters appeared where we clutched the pitchforks and we got through gallons of rich Jersey milk, far preferable to the traditional cold tea. Anyway, after turning in this way more times than I can bear to recall, the threat of rain loomed in the grey clouds building up overhead.

Rick sent an SOS to Rob for immediate help – the book said top-quality hay was cut at the flowering stage and dried rapidly. Should rain happen to fall on the winrows, this quality was lost. Afternoon and late into the evening, we went to and fro with the tractor and Scotch cart on rubbers, which Rick had converted to carry more hay by adding supports at front and rear, whipped in position with binder twine. Even so, the pitchforks of loose hay rapidly filled the cart. We were hot, dusty, irritable and worn out when the last load had been thrown off in the barn, and we felt we had earned our rest that night, happy in the knowledge our hay was safely home.

That satisfaction did not last long though. About a week later, on checking the hay for signs of heating-up (spontaneous combustion can occur if hay is stored at too high a moisture content), we discovered it was definitely too warm. The barn, the former cowshed, was not ideal as a Dutch barn, owing to poor air circulation. Rick removed the wooden slats that comprised the back of the shed to allow a through draught and we pitched the majority of the hay all out of the barn again; there was hay everywhere, in the yard, around the sheds and up the pathways. Then, that night, it rained; and it rained every day for a week – and each time it rained, we had to get out there with our forks and turn it! We were sick of that hay. It was no longer good quality hay; in fact, it now smelt positively musty. We could not face handling it anymore, so we called in George Sharp, a contractor who baled it for us, together with fourteen bales taken from the second acre that was running away with Blossom, the five cades, two sheep and the four geese. I vowed never to get involved with haymaking again.

This year, we put the haymaking completely in George Sharp's hands. He would cut, turn and bale for about thirty pence per bale with his tractor and equipment in a fraction of the time it would have taken us. Cutting took eight hours, which included one acre set aside here, Scarborough Bank (the road on which we live and now have the rental for at £25 per annum), and a two-acre field, lousy with weed, that we had been given for hay – the owner wanting it 'tidied'. The weather was kind to us for the first time this summer and no rain fell on it. George Sharp did not touch it for a week and a day – how much simpler than our effort – then rang at nine in the morning to say he required the dykeside hay tossing into two winrows close to the road's edge, as he would not be able to get the hay tedder that far down the drain, and he would be there after lunch to bale. Now this was a bit of a shock, as we had contracted him to do the lot. It really was a ridiculous request – there must have been nine miles of winrows in all down the Bank.

Anyway, we began hopelessly turning – Bryony in the baby carrier on Rick's back, Clyde trotting beside us. We had not even made the corner by 11.30 a.m. – the traditional midday break here. In the afternoon, George's son drove down with the tedder, to our top field. Then George arrived, plus baler. We were by now working steadily towards the corner on this side of the road. The two tractors were very soon upon us and then it was amazing. The hay tedder was, after all, able to reach right down the drainside and our day's hand-turning had all been in vain. At this point, I left to stack the baled hay into piles of eight, while Rick joined the tractor team. There were 569 bales in all - 94 from our acre, 375 from the roadside and 100 for free - which were sold off the field and once costs had been deducted accounted for the quickest £50 we had ever made!

All in all, we were well pleased. In the new Dutch barn that Rick had constructed from telegraph poles (£6 each from the local electricity board) were 369 bales at ten pence each and very little effort, with the roof nailed in place over them; now Rick had the stack to climb up on.

So another era has passed. In our enthusiastic early self-sufficiency days we had to do everything, but we have now matured to a realisation we cannot do it all and do it all well. If the strawberries can bring in a good income (and they probably coincide with the haymaking season too) we are content to let a machine do the hard labour for us. If it had been something we particularly enjoyed, then it would be different. Perhaps one day we may buy the necessary implements of our own - cutter bar, a tedder or a woofler and baler - then we could cut out this extra cost and just get on with it when the hay was ready, rather than wait our turn on the contractor's list.

CHAPTER 22

Winnowing was no problem

When we started out, we were going to do everything ourselves. We had already bought the Atlas hand grain mill and were grinding our own flour from wheat grown in sight of the house. We had also by now collected together an assortment of animals, so, naively, we went ahead with the campaign to 'produce all our grain requirements' (bearing in mind we were on an acre holding with half an acre rented) and broadcast a very small patch. Harvest time came round and Herb appeared through the gap in the hedge and was keen to demonstrate scything the corn, tying it into bundles, or sheaths, held together with straw in a particular knot, and then stacking the sheaths into stooks of six or eight. Twenty minutes later, he was finished. The following year we again put in an area of wheat, undersowing with a permanent pasture mix, in the first fenced enclosure that Grace, the sow, had cleared. But, in both instances, on this small scale we were only playing, the birds took such a vast percentage and the wheat, sheaths and all were tossed daily into the hen house and in a very short time were gone.

We made no further attempts until moving to Fir Tree Cottage. Having moved in on 21 March, we found the field to the west of the house under water - the area planned for barley, wheat and fodder crops. We were consequently held up in preparing the land and drilling. It was into April before Rick hitched up the fertiliser spreader to the David Brown and thus broadcast half an acre of barley and one acre of wheat. We found the spring sown barley grew rapidly and quickly got ahead of the weed, but the similarly sown wheat became swamped in it.

So harvesting came round. Rick and I spent one afternoon cutting the barley by Allen scythe - making very slow progress. The same problems showed themselves as with the hay, Rick taking only half a cut, so that the straw fell away from the cutter blade. I followed behind, scooping the straw into sheaths and tying tightly with binder twine - a slip knot and two half hitches. I found

I could achieve a far firmer knot than with the old straw bands. Periodically, I would flop down on the fresh stubble, annoyed by the gad flies and the heat, and troubled with morning sickness that lasted all day. We banged the sheaths hard into the ground, heads touching in pairs, with one either end of the tent shape so formed.

Not making much impression, Steve and Gina volunteered their labour the next afternoon, and eventually the job was done. We stood back satisfied to see the gleaming stooks standing proudly for all to view. The corn was then left in the field to hear the church bells ring for the traditional three Sundays.

The wheat was very late. Being now very much aware of the time element involved, we put the matter into Harold's capable hands. He came along with his combine harvester and rendered the grain in fifteen minutes to a state it would have taken us a week to achieve. Ironically this was in return for Rick cutting an acre of wheat for him with the Allen scythe - the Environmental Department had demanded trees be planted to conceal his poultry houses; then he realised the combines could not get in to harvest the acre of wheat there. We seized the last clump of straws left standing, where traditionally the spirit is held and where a rat, mouse or rabbit is usually crouched, before they too were pulverised. Then Sue, a friend who makes corn dollies, dampened, plaited and twisted the straws into the form of a walking stick and an umbrella, thus trapping the spirit within, to be released on breaking open to resow in the spring.

However, the grain was cut at nineteen per cent moisture, because he dare not leave it any longer in the already late fall, and as, by our organic principles, we did not use a pre-emergence spray, so we did not use Round-up to kill off the weed standing well above the spring-sown crop. The combine did not differentiate and the resulting wheat was full of weed seed. We winnowed and sieved, and sieved again, but still many seeds got through into our baking, giving our cakes a strange garlicky flavour. Because of this, and afraid of being left with a ton or more of mouldy grain on our hands due to the excessive moisture content, we kept a quantity back for our hens and sold the rest to friends and acquaintances as poultry feed. Then we went out and bought half a ton of chemically-grown milling wheat, which tasted a lot better than our organic!

Last year, we broadcast the wheat in the winter, and it got ahead

Our Land at Fir. Tree Cottage

of the weeds, further helped by undersowing with a ley in the spring; thus we solved the problem of growing corn without sprays. Again we cut and stooked by hand, and later carted the sheaths down to a stack in the yard, protected from the gales by netting, though not from the mice, hens and rats too. But as to the future of our grain harvest, we have concluded - never again.

Yes, it is very satisfying eating a sandwich and knowing you have produced it completely yourselves, but financially it is not worth the effort. With our very primitive means of harvesting a lot of grain is lost. As for the threshing, I spent a week bashing the handfuls of wheat against the side of the wheelbarrow. Fortunately in the fens, winnowing was no problem!

We have chosen to buy in our grain this year, grown in the field opposite our house, to put the spare acre down to grass and sell off a lamb or two to cover the cost of buying in. Since we took the plunge, after the pig farm where Rick worked part-time had closed down through disease, and are now reliant solely on the holding as our source of income, it means we have to really look into the economics of everything we do and we must turn to growing those things which offer us a good return - soft fruit, flowers and so on. If we can grow 6,000 lb of strawberries on half an acre, what is the sense in 'playing' with wheat? So, a little sadly perhaps, haymaking by hand, fodder crops and grain growing have been put aside as being too consuming of our precious time and economically unsound for what we strive to achieve.

For stockfeed, we have always grown kale, marrowstem (to feed when the grass declines, chosen as the cows will readily eat the stalks too) and thousand head (for it withstands hard winters); fodder beet to feed before Christmas and mangolds after - forty tons of which can apparently be pulled from an acre and lots of muck. Both have a higher food value than turnips and swedes. All these we have only fed as a supplement to hay and some concentrate, but even so, they give the milk yield a real boost and the cows relish them on their dry diet. Incidentally, we clamp the roots, wilting and feeding the tops after a few days, then grind prior to feeding in a root cutter we picked up for £25. It is said that mangolds are slightly toxic until ripe, that is why we use the lower yielding, but more nutritious fodder beet first. For two seasons, we grew surplus fodder crops for sale, having a ready market among smallholders locally, but decided a lamb would reap the same reward for nowhere near the work involved, taken care of by our increase in grass acreage.

Any waste produce is put through the animals, but nothing further is grown for the sole purpose. Pig potatoes can usually be obtained free here, or for £3-£5 a ton at the most, so they are certainly not worth the time, trouble and extra land; and if you have lost your own potatoes through your neglect in protecting from blight, you can live a winter on them too! At Wisbech, we could get hold of an unlimited supply of carrot tops too, between August and February, eaten by all the stock.

BREAD AND YEAST

We have made our own bread ever since the bakers' strike back in 1974 but I did wonder if I had gone too far in drumming its merits into the children when Dicken, returning home after his first day at school, showed his alarm at finding one of his friends had white bread sandwiches in his luncheon box. He whispered to me 'White bread makes you dead, doesn't it?'

Here is our easy loaf recipe; it is very time-saving, as at no stage is it kneaded. I do not weigh anything – it is all so variable anyway, the quantity of water it will absorb depending upon whether the wheat is hard or soft and how coarsely the flour is ground.

Bread Recipe

Into 2 pints of lukewarm water, milk or preferably butter milk, stir a dessertspoonful of sugar and shake half a packet of dried yeast onto the surface. Leave to froth for approximately 15 minutes.

Take a large mixing bowl of flour; salt to taste. Pour in the yeast and mix in well with the hands, adding further liquid as necessary until the resulting dough is on the sticky side. Leave to rise, preferably overnight, by the Rayburn if it is lit – but at room temperature it will still ferment, just at a more leisurely pace.

First thing in the morning, slop the mix into bread tins, for the size required. We use two small and one large – only because that is the size we happen to have. Bake at 450°F for 20 minutes and until easily removable from the tins. Put back in stove until done – they should sound hollow when tapped on the bottom.

We like fresh yeast best of all, but as we only shop once a fortnight and are ten miles from the town, we usually use dried yeast for convenience or else buy a block of fresh, cut it into cubes and pop it into the freezer. But the price goes on rising, so I am giving a recipe I have accumulated for home-made yeast.

Yeast Recipe

Boil 1 lb flour, ¼ lb brown sugar, little salt in two gallons of water for an hour. When the milk is warm, bottle and cork it. It will be fit for use in 24 hours. A pint of this yeast is sufficient for 18 lb bread.

Sour Dough Bread

Make a well in the centre of a bowl of flour. Pour in a mixture of warm water, sugar and salt. Leave in the warm for a day and it should ferment. Mix together and bake as usual.

CHAPTER 23
A splash of colour

Besides soft fruit and couch grass, Ivy Cottage was set out with perennials for the cut flower market - irises, Esther Reads (Corban Gold and Wirral Supreme) pyrethrums, Sarah Bernhardt peonies and also Early Messenger sweet williams, treated as a biennial - sown the first year, pulled and ploughed in the second. When we moved in at the close of June, all but the Esther Reads were over. Maud was heavily tied up with the strawberries, and as the flowers bring in a trivial amount by comparison, she allowed us to market them. So, with her still present, we learned the trade as we went along.

Flower picking looks a romantic occupation to the passer-by, especially if the sun is shining down on you, but in reality there are usually many more tasks to accomplish before nightfall, so we are not working at that pleasant leisurely pace and the hot sun is wilting the flowers still in your hand. It is a case of hurrying back and forth with bundles to what shade there is. The sun does not always beam down either, many is the day we have pulled in the rain to complete an order soon to be collected or to make up a box for market.

We learned how many heads to put to the bunch - ten Esther Reads and pyrethrums, five irises, a bouquet of unfixed quantity with sweet williams, and single blooms in the case of the magnificent peony; and how many bunches to the box - thirty Esther Reads, irises, pyrethrums and sixteen to twenty of the bulkier sweet williams. The bunches were held in place by stays, a half inch wide strip of recycled metal with pointed ends, that are pushed through the sides of the cardboard box, so that the flowers were not tossed around in transit.

They were gathered the day previous to haulage and placed in galvanised flower buckets in an inch of water. Before packing, the bunches would be drained for about half an hour maybe, so as not to dampen the box, and, in this way, they would be fresh and attractive after their journey, in readiness for the auction on the following day. Thus a minimum of two days would elapse from picking until the blooms appeared in the shops. At weekends and bank holidays, the time lapse would be even greater - picking Saturday or Sunday (as the flowers can be brought out too far), haulage Monday for the Tuesday market. All our flowers from Walpole went up to Smithfield Market, Manchester.

As with all things, we have had our ups and downs. The best year for flowers was 1977, the Silver Jubilee, celebrated in early June when the bulk of our perennials were ready; and the most disastrous was the time we sent one box of the first pyrethrums - taking into account the price of box, rubber bands, stays, transport, commission and porterage - and received ten pence. Not much to pay for our labour throughout the year - keeping the plants free of weeds, picking and packing. Also, our first attempt at sweet william growing was a disaster, because they coincided with Rick's first attempt at push-hoeing. Borrowing Herb's wheeled implement, he proceeded to walk up the row, destroying all our tenderly sprouting plants rather than the carpet of weed seedlings in the alleyways between!

Besides weeding, very little work was involved with the flowers during the remainder of the year. As with all things, a feed occasionally is appreciated. The dead pyrethrums and Esther Read leaves are mown off after flowering, together with any unmarketable heads, the inevitably diseased and deformed ones - all is not perfect in nature. As the peonies grew up in the spring, all the secondary buds had to be nipped out to allow only one main central bud to blossom. The peonies and irises were gathered when only a tip of colour showed in the bud; to hold any length of time, they were then laid out on a concrete floor. The night previous to haulage, they would be stood in water. Pyrethrums and Esther Reads are picked when fully open. We were taught to cut the stalks generally as low as possible, but the irises were pulled, revealing inches of blanched extremity.

Behind the greenhouses, clumps of chrysanthemums grew in the mature flower borders, amongst assorted shrubs, evergreens and cottage garden flowers. From these, Maud would take cuttings, dipping them in hormone powder and planting them in seed trays, for setting out in the glasshouses when the tomatoes were done. But the chrysanthemums seemed such a specialised area, and the root-stock well past its best, that we chose not to dabble in that field.

In our attack on the underlying couch, the flowers were gradually lifted, the roots and bulbs split up and sold and more strawberries took their place, but we had enjoyed the splash of colour they brought us. Motoring past the rose fields at Walsoken, near Wisbech, and through the bulb areas around Spalding, I do not cease to be delighted with the acres of beauty. We had found them resilient to all attack and a little welcome cash before the soft fruit brought our bulk income.

At Fir Tree Cottage, we have to date only sown sweet williams. Drilled at the beginning of May, by August the plants were bushy and of a good height, but the leaves appeared strangely furled. Fir Tree Cottage being a registered holding, gave us entitlement to utilise the advisory services of the Ministry of Agriculture free of charge. Therefore we called ADAS in, who collected specimens of the plants and took away bagfuls of our precious soil for analysis, to exclude a deficiency. The results came back in due course – it was thrips from which the sweet williams were suffering, thunderflies, the grain pests. Incidentally, the pH of that area, the north-west acre now under grass, was 6.1. However, we need not have bothered; the block lay under water during the flooding the following April, and at flowering the leaves were so mottled and yellow as to make them unsaleable. We ploughed them in.

Undaunted though, we drilled a further block that spring, next to the road to the west of the house. The plants stayed healthy; they took a battering during a spell of heavy rain in the early spring of the second year, but the sunshine gradually lifted their heads up again. We had a buyer for them who marketed in Boston, Grimsby and Cambridge, importing mainly from Holland, but going world-wide for flowers.

Then, just before the first colour showed, the plants suffered an attack of aphids. No one would buy greenfly-riddled flowers, so being organic we began to spray the crop with a loaded knapsack sprayer by recommended organic means. First we tried washing-up liquid – which had no effect whatsoever, then some disinfectant we use for the pigsheds – but they were still there in force; to Jeyes fluid – no use; then some liquid derris we found on our shelves, but still it did not kill the pests. We therefore took our troubles and organic principles to a large seed-merchant in Boston, who suggested nicotine spray and although expensive, it did not work. By now the blooms would soon be ready for picking, we were busy

with other things and, needless to say, were finding the whole matter extremely irksome. So, we went ahead and bought a bottle of dreaded Malathion and proceeded to sell over 2,000 bunches. At the beginning, we were instructed to gather, without using knives, all heads with any colour showing, for which we received fifteen pence a bunch. As more flooded into the market, this dropped to twelve pence and eventually to eight pence. The heat wave in progress opened the heads behind us; the spectrum of colour shades - such a wealth of old-fashioned beauty - was a wonderful sight. Peggy and Eddy, the first caravan guests, helped to gather in the luxury crop, with tins fastened around their waists by binder twine. We lay the bunches out upon the concrete floor as requested, where they could remain without harm for several days.

After the latest disastrous summer, we are considering stocking an area here with perennial flowers, because perennial crops cut down the work - drilling, controlling the weeds when those seedlings are small and so on; also, a permanent crop can be controlled easily by organic means - a black polythene covering. This would be uneconomical on an annual crop, where if the plot size varied even by inches the following year, the polythene would become obsolete. Provided a local market can be ascertained, flowers will reap an income before the other crops are mature. It is then that all our bills fall due - rates, electricity, telephone, insurance, car tax, the list is endless - and we are desperate! Flowers also seem fairly resilient to attack, and some even have a beneficial effect on other crops: French marigolds with their strong odour ward off white fly; wormwood attracts the blackfly away from the broad beans; mint hung in the porchway of the house, draws the flies to it, so they will not cross the threshold. I would like to experiment with pyrethrum plants, from which a strong insecticide is extracted, to discover any effects on pests of intercropping. But, above all, their beauty brightens up our lives a little.

CHAPTER 24

The fruits of our labour

At Ivy Cottage, soft fruit, cut flowers and the produce of the greenhouses were the only crops grown specifically for sale. However any surplus of the large variety of vegetables we had for the house, would be marketed. In fact over £600 profit from our crops, excluding animal products - milk, eggs, kids, weaners - was usual on that one acre.

When we moved in, and during our five years there, we were fervent organic gardeners. We would muck heavily, bringing in from outside in the early days (bullock manure from Len Powers, duck litter for the allotment) but very soon our own stock gave plenty; the fertility coming in from roadside grazing, the verge-mown grass, hay, straw and purchased concentrates, food scraps and vegetable waste from the gardens nearby, even lawn clippings brought home from Rick's gardening job - really anything that could be scrounged for the compost heap or to put through the animals. By and by, the soil became good and rich; the fine plants and their subsequent yields were the proof. In time, the earth was well-mulched, which inhibited weed growth dramatically.

It was a simple matter to let nature take its course. Our living was not dependent upon our crops, and so we won some and lost some. One year, after an invasion of cabbage whites, we puffed derris dust over the riddled leaves, but we never bothered again - the brassicas always outgrow the attack. The wrong attitude perhaps on a saleable crop for no one will buy cabbages that do not look nice, but we did not mind sharing a little of our own dinner with the slugs and snails or hodnedods as the locals called them. The gooseberry bushes were carpeted by a lush growth of couch grass, where an army of caterpillars lived. One June day, they moved in and very soon there was not a leaf to be seen. We were pleased to be able to easily distinguish the fruit, normally so well-disguised amongst the foliage, and in the autumn, the plants came to life again with fresh green clothing.

The soil type at Stickford is different, by far the worst type for growing the weeds, so much a problem to us now, that we could possibly have chosen. We still started out here staunchly anti-chemical - our decision to compromise came because we have not been able to reap a reasonable income from these seven acres by our ideals, and we are anxious to relieve the great pressure we always find ourselves under. We enjoy the hard physical work and the feeling of contented exhaustion at the end of the day; but not the continual hurrying here, there and everywhere, with no time to stop and enjoy the fruits of our labour, and the constant nagging of the weeds growing up behind us.

It was not until we were involved with our crops on this larger scale that we became aware of the farmers' problems. How hard we have tried to sell the quarter acre of beetroot and spring onions, that are getting more and more unsaleable as the days go by - the market is flooded, the price too low. My sympathies lie with the poor farmer in this glut.

When I was at school, summer was when we broke up for the long vacation and went away for two weeks' holiday, but I have come to realise that summer is really over then. The height of the season is the equinox, Midsummer's Day. Harvesting our produce begins then, bare patches slowly appear and, by August, the crops surrounding us - barley, wheat, potatoes and beet - are gradually gathered in and the land is ploughed, changing from gold or green to brown, blending with the autumn colours of the changing trees.

In June and July, spare moments are spent podding, slicing, washing, blanching; preparing peas, french beans, runner beans, small carrots, tomatoes (so delicious with home-cured bacon), and a special pack of small new potatoes complete with a sprig of mint for Christmas dinner, all for freezing. If the cauliflowers and sweet corn do well, they will follow later. Carrots are pulled at the beginning of September, before a secondary invasion of carrot fly, and will keep well in sacks until about February. Potatoes we clamp, alongside the roots for the stock. Onions and shallots are stored outside, sheltered from the rain, unaffected by the frost. Likewise, the swedes stay in the ground until needed.

Before we acquired the freezer, I learned the technique of bottling and so preserved apples, blackberries, strawberries and pears in brandy for special occasions, and also our vast crops of plums from the two Victorias that stood on the headland, bordering Ivy Cottage and Saddlers Piece, which Audrey had purchased along with the acre and a half in which Rosedene already stood. However, it was more trouble so now we just pop them into the freezer, but I do miss those groaning shelves. One cannot show off a full freezer in the same way as a larder of bottles, jams and pickles. Our family is not very fond of the latter, but at one time or another I have tried chutneys, piccalilli, beetroot and samphire in vinegar, pickled eggs, onions and red cabbage. I am including here a recipe for home-made vinegar, which is certainly a saving on bought vinegar when pickling in quantity.

VINEGAR RECIPE

1 pint acetic acid	a little burnt sugar
1 gallon water (boiled)	

Method Pour water over the acid and sugar. Allow to cool. Bottle.

Mint sauce is a must for our two autumn killed lambs. Salted sausage skins, salted runner beans, and two crocks of eggs, preserved in water glass, complete our shelves, with sacks of wheat off the ground on wood beneath; comforting reassurance against the lean months of winter ahead.

OUR MENU OF VEGETABLES THROUGHOUT THE YEAR

January Carrots and onions from sacks, and the following from the ground: Brussels sprouts, January Kings, purple sprouting broccoli, Jerusalem artichokes, swedes, celery and celeriac.
February As above.
March Brussels sprouts, January Kings, purple sprouting broccoli and onions.
April Brussels sprouts, January Kings, purple sprouting broccoli, onions, but also starting on the frozen vegetables.
May Basically frozen vegetables – tomatoes, carrots, peas, french beans and runner beans. Spring cabbage as ready, cauliflowers and white Lisbon spring onions if overwintered successfully. Chives, Welsh onion tops and other perennial fresh herbs which get off to an early start, give that first green touch to our by now indifferent diet, and can garnish sandwiches too.
June Frozen vegetables and fresh herbs.

Midsummer's Day At this point the cycle renews itself and the following are picked young and are a welcome addition to the table: new potatoes, broad beans, carrots, peas, radish, spring onions, spinach, golden acre cabbage, celtuce, cress, French beans and beetroot.
July Runner beans, tomatoes, cucumbers, sweet corn, vegetable marrows, spaghetti marrows, samphire, shallots and other vegetables as in June.
August As above.
September Marrows, cauliflowers, tomatoes, onions, carrots, artichokes, spinach, cucumbers, beetroot, (celtuce, lettuce, cress, radishes - if we have kept up the successional sowing).
October Brussels sprouts, tomatoes, onions, cauliflowers, marrows, artichokes, swedes, carrots and beetroot.
November Tomatoes, January Kings, Brussels sprouts, onions, swedes, celery, celeriac, artichokes and carrots.
December January Kings, Brussels sprouts, onions, swedes, celery, celeriac, artichokes and carrots.

The above does not include any vegetables we might produce out of season in the greenhouse.

Swedes and dried beans are only eaten on winter days when the Rayburn is in use, as they are very energy consuming. We are not very fond of turnips, leeks, parsnips and winter radish, so some years we grow them and other years we do not. We always put in a few early potatoes, to get sweet new roots to have cooked with a sprig of mint and served with lots of butter; these are ready soon after the longest day, before the main ones, which crop more heavily, are ready. One year, forgetting all about them, Rick ran the harrows over the area and it was not until he spotted pieces of potato everywhere, that he realised what had happened to his carefully hand-dug trenches. Each year we try another unusual vegetable or two, and now we are introducing calabrese and everlasting cauliflowers as a permanent addition to our diet.

When we have too many cauliflowers, a favourite dish of mine is to dip cooked cauliflower florets in a batter mixture and fry them until brown.

THE GREENHOUSES

We inherited three commercial glasshouses at Walpole, though they were a mixed blessing. Two of them measured fifty-six feet by fourteen feet, the other twenty-five feet by fourteen and they were included in the purchase price of £6,000 for property and land. Our primary intention was to make an income from the glasshouses. Each year we planted up two houses with tomatoes around 20 April. The third house varied; we tried French beans, which fetched forty-five pence a pound home, and cucumbers commercially. Also, we discovered that sweet corn fared excellently in the houses whereas outside it was prone to failure – experiencing only two decent seasons out of nine; the cobs were variable in size, but we only grow them for the family, with whom they are very popular. As another luxury just for the house, we sow a few French beans in July and August to provide a taste of tender young vegetables before the less interesting brassicas and winter roots are all that remain fresh. We did attempt overwintering lettuces, but they fell prey to aphids and decay. Heating costs were prohibitive and so the lettuces did not mature in the middle of winter. The price received for the effort barely covered the cost of the plants in peat blocks, the boxes or the necessary polythene bags. The rabbits, at least, benefited from our mistake. From then on, one house doubled up as a Dutch barn in the winter, and a second as winter quarters for the hens, which relieved the build-up of worms in the hen house.

Sayings abound about greenhouse cultivation – which, if any, are just old wives' tales I do not know; for example, you must not plant potatoes near a greenhouse; you must not mix tomatoes and cucumbers in the same house; and spring onions will not grow under glass.

At Fir Tree Cottage, there is only a garden-sized greenhouse, composed of twenty Dutchlights, plus ends, so to date we have only produced tomatoes, cucumbers, melons and sweet corn to supply the house. In full swing, it looks really productive. I also enjoy the greenhouse in May, with seed trays everywhere, awaiting the abatement of the frost – various marrows, pumpkins, ridge cucumbers, runner beans, sweet corn, some flowers – such promise of the autumn fruits.

Next year we intend to invest in a polytunnel to provide vegetables for sale to the 'pick your own' customers. Basically we will provide cucumbers and tomatoes, the latter being a favourite of mine. I enjoy watching their rapid growth, spraying them with water to set the trusses, nipping out the side shoots, removing the lower leaves to ripen the fruit; we count on eating our first cheese and tomato sandwich by mid-July, without heat. At the end of October we gather the remaining green tomatoes into the house, where they will still ripen gradually.

Then the fattening cockerels that have been born to free-range are confined in the small house, where they will fill out on barley meal and eat the bugs. They are deep-littered on straw and all the waste vegetation accumulated as we tidy the holding in preparation for its winter dormancy, so they also fertilise the soil, making a soil-change unnecessary.

Incidentally, excellent containers for use in the greenhouse, can be made from old fertiliser bags. The bags are cut into strips three inches wide and stapled to form bands of about two and a half inches in diameter. The resulting pots can be stood on potato trays and filled with peat. They take up far less room than the traditional plant pot and are excellent for pricking out seedlings before finally transplanting into the ground.

CROPS FOR FREE

Samphire is a crop of the sea, which along with the occasional eel or the unexpected find of a clump of field mushrooms, is an added free bonus which can be found in the Marshes that extend around the coastline from Skegness to Sutton Bridge. On our first trip, we were looking for seaweed growing waist-high, until we realised it was another of Herb's stories! Care has to be taken in wading out on the mud-flats, both to ensure one does not sink past the knees in mud, and also that the tide does not creep in, stranding you – many lives have been lost in this way. We would often walk out past the lighthouse, Peter Scott's refuge in the war, gathering marsh statice along the banks where cows graze. At Gibraltar Point we would fill our bags with the bright green weed, resembling tiny asparagus shoots, and linger awhile in the dunes, with the sea miles out – the scene of past summer solstice celebrations; charred remains of camp fires are evidence of similar night parties. A healthy appetite acquired, together with the drowsiness the sea air always seems to foster, we went home to prepare the delicacy - a wash to remove excess salt, the roots snipped off and five minutes in boiling water. Served with plenty of butter and a liberal peppering, it is a 'fun' food, eaten with the fingers and pulled through the teeth.

Puff balls too were sometimes to be found in the waste areas around the Marsh - their name arising from the cloud of dust that flies up when the balls (sometimes the size of dinner plates) are kicked. If they reach this state, they have gone too far, but when young and fresh they are delicious sliced, fried in butter and eaten on toast.

Other such exploits we have tried are crabbing, when I slipped off the rock and received a soaking fully clothed and fairly pregnant. In any case, we could not distinguish between the edible crabs and the inedible ones! We also went cockle-hunting at Heacham - poking around in the sand soon after the tide had ebbed, only taking those shells that were still firmly clamped together.

With all these things, the blackberrying rambles along country lanes and the beach trips, the pleasure is not only in the eating, but in the actual expedition and the satisfaction of gleaning by one's one initiative, food for free.

* * *

We have at last acquired a lawnmower and I dream one day of creating an old-fashioned cottage garden. In the meantime, we had an unusual visitor, clutching to a flower stem, tapping at a seed pod with his beak; the reference book identified him for us as a goldfinch - a bird of waste land!

CHAPTER 25

Odd jobbing

Purchasing Fir Tree Cottage devoured all our spare cash, so an income was necessary straight away. Rick put an advertisement in the village shop and at the local feed mill as an odd job man. This received a very varied response, but no realistic wage. After three weeks at Stickford, a part-time pigman's job was advertised in the local paper, within walking distance of Fir Tree Cottage. It was a two hundred sow unit with one other worker, but the owner was reluctant to employ Rick, from the disease angle, as we kept pigs ourselves. However, by the end of the weekend, he had conceded and Rick embarked on his 8 a.m. to noon job, that was to become a permanent fixture.

The job was unbelievably convenient - a regular income, within minutes of the house. Rick would spend the mornings working steadily there, and the afternoons doing what he liked here with his family - and so we moved in a pleasant rut. With the introduction of the job creation scheme however, the work became more and more part time, first losing half an hour a day, followed by a reduction to only three mornings a week. But we were slowly becoming more established here and we strove to survive on this very small income.

Then disaster struck. The weaners in the following-on pens (eight to twelve weeks) started dying. Only an odd one now and again at the beginning, put down to bowel oedema or some similar cause, but then it began occurring more regularly and the Ministry were called in. Aujeskies disease was suspected, bodies were sent for analysis and it was confirmed. By now, some of the sows were also sick and the boars too, so the sows were not being served. The sick sows that were due to farrow had to be moved out of the open air sow yard into the farrowing houses - and thus the disease spread more and more. It was just impossible to contain, with the system as it was, a vicious circle of infection. Tighter restrictions were immediately brought in, but it was shutting the stable door after the horse had bolted. Rick was laid off, for financial reasons, until the situation improved. He went back for a while just after Christmas, but things were no better. After a while, we began to notice less and less pigs there and heard rumour as one does in a village. The premises were to be left devoid of all swine for a certain period and made sterile. All sows and gilts had been sold.

When Rick's part-time job at the pig farm folded completely, strawberries and the other fruits of our labours were looking promising, so we saw this as the final step to our complete independence - from now on we would rely solely on the holding to provide our daily needs. But then fate struck again; the strawberries failed and consequently the other crops did not sell well. This was the end of our eight years of organic husbandry. We also arrived at a cross-roads in our lives: we could let our depression take over and throw everything up, we could plod along the same route never really getting anywhere, or we could fight - change course completely. This is the path we chose. But if we were to carry on surviving here, money had to be earned to see us through the winter, so Rick started to look around.

Completely by chance, Roy telephoned asking if Rick would construct a dog run for two alsations, which in turn led to two further days employed installing a shower unit. Then, Rick was lucky enough to receive an offer of several weeks' harvest work from Allewells Farms Ltd - a large concern stationed in the village, but with two thousand acres all over the place. Only six men were employed full time, which left things a bit tight at this season of the year.

And so began several weeks hectic beyond belief for us. Getting the grain in is of prime importance to the farmer; the men are expected to work every hour that the moisture content is at the correct level; at 7.30 a.m. they reported to the yard and by 9.30 a.m. the procession of combines could be seen making its way along the roads to the next field. Breaks were as short as possible; half an hour for lunch, quarter of an hour for tea at four in the afternoon. But there was a great feeling of community spirit in the throng of activity, all totally involved in bringing the harvest home. The combines churned with ease through twenty acres of standing corn in a day, working on into the evening, when the beam of headlights lit up the task in hand.

Rick would stumble in, ashen faced, just after nine, and flop into his special armchair - only to pick over his meal, past hunger, too

tired for conversation. More often than not, we would just have a milk drink and retire to bed, ready to begin all over again in the morning. Not only was it a thirteen and a half hour working day, but it was for seven days a week. Rick only had one day off in those first five weeks. We thought it worthwhile when after the first three weeks, we were able to completely pay off the second-hand car we had had to buy when the old Escort van gave up.

It was pandemonium for me on my own here with the needs of four children to administer to, caravan visitors to entertain, the house to run - with bread and cheese-making, meals to cook, all the livestock to be cared for, including roving sheep and stray heifers, the evening milking, and first and foremost our own particular harvest - onions, carrots, potatoes, mangolds - and the war on weeds. Rick would milk Blossom early in the morning and that was the extent of his participation in the smallholding those first five weeks, but it had to be, for we were desperate for the money.

I thought I was coping miraculously - the possibility of getting 'all worked up' by the time Rick finished at Allewells looked like becoming a reality and then the letter came: forty-eight hours' notice that I was to be admitted to hospital for hernia repair, a matter of some urgency for which I had been waiting weeks! The immediate reaction was of panic and despair - so many things would now have to be shelved or just would not get done, and how would we all cope? Then practicality set in; Rick's sister came to look after the children, which made it possible for Rick to carry on working, albeit a shorter day. The harvest was onions and potatoes now and the pressure was off. We got through somehow, with the odd crisis or two, and I was only in hospital for five days.

Jade had the following week off school to manage things when I first came home, which he did admirably for his ten years, and in acknowledgement was awarded his home help badge at cubs. He would get Dicken off to school, look after Bryony and Clyde during the day, make bread, butter and cheese, cook a snack at lunchtime, prepare vegetables, sweep up, wash up, use the washing-machine and tend the animals. I slowly got back on my feet again and, at the same time, the weather worsened. As the rain set in, Rick found himself tediously stamping potato bags and so his work ended; but we had weathered another storm.

CHAPTER 26
Working holidays

While we were living at Walpole Marsh, we got to know John and Marilyn, who had a five acre smallholding at Walpole Highway. John was a full-time solicitor and kept his land immaculate by rising at an unearthly hour and retiring late. Marilyn enjoyed being the 'farmer's wife' and thrived on the inside chores. She was an excellent tailer and made all of John's suits and shirts for the office, and she tended the goat herd, freezing the milk for sale. They ran two Reliant motor cars and lived with the minimum of modern devices, without hoover, washing-machine, or television. They were vegetarians, for economical reasons rather than conscience, although the holding was mixed (a sow, goats, sheep, geese, hens) and run with a view to making a profit.

'Kenilworth' was always full, with their friends, foster children – their own child, Thomas, was born during this period – or working guests. John would advertise for these helpers, on the basis of food and accommodation in return for help. Some would spend part of a day working with us at Ivy Cottage. They came from a wide range of backgrounds and their reasons for replying to John's advertisement were also varied.

When we moved to Stickford, we thought we had enough experience, work and room, to be able to organise 'working holidays' of our own. William Brighouse-Smythe was our first – an agricultural student seeking work experience. He participated in our pioneering work here – planting our first strawberry block, fencing paddocks and then drilling the root break, with Planet seed drill. I think it was with some relief that he escaped home, away from our primitive living conditions, a cold house and the bleak bare plot.

It was late summer before we actually placed an advertisement in *Practical Self-sufficiency* magazine. Jos and Phil, students from Leicester, came to us over August bank holiday weekend, and with four pairs of hands we attacked the weeds in force. Jos and Phil were limited to living in the city by lack of finance, but did what they could by growing as many vegetables as is possible in a tiny city garden, made good wine (which they brought for us to sample) and baked their own bread.

Next were Jon, Rosanna and Gwen (who, incidentally, was a well-behaved dog). They were an energetic young couple, hoping to have a smallholding of their own one day and they already had a foot in the door by having bought and renovated a dilapidated old house. They rose at dawn and had the stock fed and watered before we had even thought about it! Throughout their stay they worked extremely hard and with interest and enthusiasm in all that they did, and were a great help to us, at a time when Rick was working twelve hours a day at the pig farm and I was in the early stages of pregnancy. Rosanna had an allergy to dust, so it was unfortunate that it was this week Rick brought home the wheat straw.

Sylvie wrote 'I have a year old child and we are living in Islington. I am fed up with being in the city and am trying to find a way to escape occasionally. When I saw your advertisement I thought it might suit me'. We were blessed with glorious September days while Sylvie and Jake were with us, and began harvesting our main crop potatoes and turnips (the roots to store for winter fodder for the rabbits, the tops to feed them there and then). Jake was a demanding child, just toddling, and Sylvia was prepared to lavish him with her full attention every minute of every hour. She had little knowledge of country matters and no interest in learning any, and only worked half-heartedly.

Roger spent the worst weather of October here - torrential downpours gave way only to heavy showers. There was mud everywhere, and most of the chores planned had to be abandoned - harvesting the mangolds and fodder beets, lifting the turnips. Even taking kale down to the cattle, prematurely brought in for the winter, was a mammoth task. We had foolishly drilled the kale in the farthest acre from the cowshed. The wheelbarrow would only move a couple of feet before the wheel became too encased with mud to budge an inch further. Words would be muttered, the wheel freed by hand, the mud transferred to the handles of the barrow and a couple more feet of progress made before the whole procedure had to be repeated. It was mainly cleaning pens and mucking out during that spell, then a dash indoors to warm up, dry out wet clothing and talk by the fireside. But we were the richer for the experience; Roger was an interesting person, he was doing postgraduate research at Exeter University on Beduins and, having already travelled and lived among the peoples of Saudi Arabia, he intended to return there for further research soon. Roger's particular interest was in sheep and goats. His dream was to be a sheep farmer.

Although we thoroughly enjoyed the varied company and being in touch again - one tends to stagnate out in the country - all in all we decided against doing working holidays as such again. I resented having to spend so much time cooking, as I felt the helpers deserved more than the makeshift meals which we occasionally have when we are busy or I do not feel like preparing a large meal. I did not like having to spend longer over household jobs either. I also found I was not a good boss. I became too impatient when someone spent all day cleaning out the rabbit hutches - a job that took me an hour at most, even when heavily pregnant; and when we all worked together, we found we slowed to the pace of the learner. Writing this two years later, I do not think this was such a bad thing - maybe we were just too earnest then!

We thought we would still like to encourage visitors to come to the holding - a caravan being the ideal solution. It would ease the strain on relationships within the family to have a little privacy, and a small charge would sort out those genuinely interested from those after a free holiday.

CHAPTER 27
'Will fell trees'

While we were living in Wisbech, we would often come across a fruit orchard that was being grubbed out, either due to a change in farm policy or because the EEC were, at that time, urging farmers to grow Golden Delicious apples. One was encouraged to go in and cut and remove the wood, which we did as best we could with only a bow saw and axe, working alongside the deafening, ear-splitting scream of chain saws. One such orchard was fifty acres of apple trees to the side of our land, that had been trees since the oldest Walpole inhabitants could remember. When removed, the wind whipped through the new fifty acre wheat field where weeds would be a problem even to the chemical farmer for many years to come.

Ivy Cottage had two open fires set in modern tiled surrounds. These were ripped out when the house was renovated. A friend bartered a wood-burning stove he had constructed from a piece of tubular steel (the sort filled with cement, used in the building of piers and moorings) in exchange for two weaner pigs. This was installed in the sitting-room in front of a simple brick fireplace with tiled mantelshelf. The gaping hole left where the second fireplace had been, in the living-room, must have been five feet to the lintel. A traditional stove had probably stood here. We picked up an Esse stove from an advertisement in the local paper at a bargain price; it was superb, with three ovens, a toasting compartment, towel rail and plate rack. It took three men six hours to move the stove on runners onto the lorry, and six men, even after dismantling every removable part possible, to shift it into position at our end. This stove was the centre of our home and the family tended to accumulate around her; she heated the rooms, provided a permanently ready oven to cook with, boiled the pig potatoes, dried out our clothes, warmed gloves and boots in winter, and was the refuge to return to after a hard day's work. The stove ran on the minimum of phurnacite, with which we stoked her early in the morning and again before retiring to bed. Thus we were able to keep the woodburner going on the apple wood or on fallen or dead poplar branches sometimes dragged home from walks.

It was a different matter when we moved from the fruit growing and market garden area to one of extensive fields of wheat, barley, potatoes and sugar beet. Lincolnshire is particularly noted for its absence of trees and hedges. Once again, the house had two modern fireplaces. While bargaining for two gilts during the first autumn at Stickford, Norma Pearce mentioned she was replacing her present Rayburn with a new model, so we snapped it up. Once installed, the Rayburn ran fairly economically that first winter on logs bought locally - probably because the wood was wet elm and hardly burned at all! Elm smoulders like churchyard mould, as the saying goes. We gave paper logs a trial, but our hand-rolling method was far from satisfactory. Anyway, with both an open fire and the Rayburn, some reliable source of economical fuel had to be found before we could face another winter here.

It was the following October that we found ourselves in the rare but happy position of having a little spare cash. I was hesitant, but Rick reckoned if we invested this money in a chain saw, it would pay for itself in no time at all. Rick first placed an advertisement in the local paper: *will fell trees, in return for the timber.* The response was unbelievable; in a short time he had fifty trees, and the list went on growing and growing. There were oaks, horse chestnuts, willows, sycamores, silver birches, fir trees, limes, elms, fruit trees and even a yew.

Rick asked Bill, a fellow smallholder, if he would like to come into partnership with him, and so the comedy tree felling team was established. Bill was quietly confident, and Rick was scared of heights! Rick began by taking down a straightforward apple tree in a small garden in town; gaining confidence, he set off the next afternoon to fell a second apple tree in Boston. It stood twenty feet high and its branches hung over four gardens, four fences and had become entwined around a metal bar. Rick mumbled something about coming back tomorrow with his mate and fled!

Bill said they should drive round to survey the trees, in order to acquaint themselves with any snags and drawbacks before embarking on them. The most promising deal was at the local lawn tennis club - two thirty feet ash, two sixty feet poplars, two thirty feet sycamores and eleven thirty feet Scots pines. Bill's first tree came down perfectly, and even Rick's first one was fine - Bill showing him where to make the cuts. His second went off course and crashed down into the tennis court.

So it went on, from job to job, disaster to disaster. One job was a dead elm, fifteen feet away from a garage full of very old, very valuable bicycles, with power lines all around. Bill had cut his wedge and decided upon where to make his incision. He was half-way through the trunk, when in his slow voice he mumbled: 'I don't like the look of this'. The tree was hollow and had started to roll alarmingly this way and that. Once eventually down safely, hours went into logging it - sawing through compost can be difficult. An old age pensioner asked to buy the wood, but even so, for less than fifty pence per hour, it had been a lot of work.

Another notable tree was an oak. The two tree fellers had taken down six oak saplings with about nine inch trunks in Woodhall Spa, and, as often happened, the lady next door enquired if they could fell an oak for her. It was a big tree and involved a considerable amount of work, tying up various branches prior to dropping them, to prevent damage to an elaborate fence below. A sum of £50 was agreed upon to cover expenses. When work was well under way, the owner's son turned up - a professional tree surgeon - and disclosed that he had refused to fell it for any less than £300! The following day was a Saturday, when Rick and Bill generally caught up on any pressing jobs on their particular holdings. Even after having such a cut-price bill though, the lady phoned up irate, because the trunk had not been carted away that Saturday. On the Monday, Rick and Bill set off early with the 'Dodge' (a World War II American General's staff jeep) and Steve's single-axle flat tractor trailer without sides, to collect the oak trunk, transport it to Horncastle and make the return journey in daylight. The trunk wobbled and shook precariously along behind the Dodge and thus forty-four cubic feet of oak was delivered to the saw mill.

The cold winter of 1981-82 was now upon us and so the tree felling expeditions became less frequent and finally dwindled to a halt.

CHAPTER 28
Welsh interlude

One February Rick suggested we went on holiday. All the usual 'buts and can'ts' were voiced and then we really started thinking – there was nothing particularly pressing to do here, just muck carting. The animals were in a dormant spell – the sheep and the sow were pregnant, but not due for another month; there were no young rabbits, the hens were free-ranging and quite independent, Cello was suckling and the other calves needed little attention. But there was Blossom – she needed milking morning and evening – but there again, she was well on in her lactation, perhaps a way round the problem could be found. The idea had taken root, especially after we saw an advertisement in that month's *Practical Self-Sufficiency* magazine. *West Wales – modern six-berth caravan to let on secluded four acre organic smallholding, situated in beautiful countryside only eight miles from beaches. Ideal touring centre, good walking, riding, fishing, shooting, available locally. Rates October to March, £10 a week.*

During the severe cold weather of December and January, Pam and John had spent a month in Kenya. We had fed their sow and poultry twice daily in freezing and difficult conditions. So it came about that Pam seemed willing enough to volunteer John to do the first milking of the day, and she would check all was well at lunchtime. Rob and Sandra were bribed with the prospect of milk (their goats were dry) and Jersey cream in exchange for the evening rounds. A week later – after cramming into the poor old van several changes of clothes each, piles of nappies, secret parcels for Clyde (it would be his birthday while we were away), lots of food and four excited children – we were off. It was to be the first time we had been away on holiday for seven and a half years, and we were in a light-hearted mood. After a long journey and some difficulty in finding the place when we arrived after dark, we were greeted by steaming mugs of coffee and cookies.

We crammed so much into those next seven precious days; we toured the weaving mills and craft workshops, bought gifts from a pottery on a forty acre commune. We visited the Alternative

Technology Centre at Machynlleth - busy even on a dull February afternoon. The gardens were bare and desolate, but there was still much to impress - the beauty of the animal buildings in natural local materials; warm solar-heated water from a tap on the hillside; the achievement of growing food in a slate mine; the recycling of human waste from even the public toilets, together with recycled toilet paper; and, not least, the sheer magnitude of the surrounding mountains of Snowdonia.

The children enjoyed excursions to the coast; collecting shells, digging holes, building dams and all that children generally do on a trip to the seaside - except this was winter and everywhere was deserted. The first day torrential rain caused rivers to bulge and turned the lane into a cascading waterfall, but from then on the weather was kind. Frosty mornings gave way to brilliant midday sunshine and we discarded our coats. We would saunter slowly up the track to collect the daily milk direct from a Friesian herd along the road, enjoying the countryside and taking life at a slow pace.

The primitiveness of the daily chores held a rustic charm - stumbling down the slate slope, beneath ash and oak saplings, with wild garlic under foot, to fetch water from the spring. Such pleasant labour when time does not matter; but it may be different on a hot humid summer afternoon, with languid flies hovering around, the stock forever thirsty, and the spring almost motionless; or, in winter, struggling through chest-high snow with frozen fingers and many layers of clothes. There was no piped water to operate washing-machines or even taps, just buckets and buckets always to be fetched and the chemical toilet to be emptied when the ground is frozen hard. Electricity had only come to Penycraig a year previously - until then the Rayburn and oil lamps provided heat and light. The authorities have been promising mains water for some time, but even so, some inhabitants on the hill have declined it.

Life is so much tougher for the Welsh smallholder than for us in the windswept fens. Timber is there for the taking, a lifetime of free fuel, but as well as the lack of amenities, life is a constant battle with the elements and predators; a battle for survival against excessive rainfall, stoats, buzzards, crows which will attack newborn lambs, foxes which steal poultry in broad daylight, and not least the soil - essentially sheep country and the minimum to the acre at that.

The week passed; snowy flurries fell during the final evening as we all drank homebrew and told tales around the table, all six children, by now firm friends, playing noisily. It was early next morning when we said our goodbyes with some regret, but refreshed and keen to work our own land again. The van chugged into life; it had contended with all the hills, though maybe with some effort and smoke. Each excursion we made, we wondered if it would bring us back again; but now it pulled slowly through the mountains, journeying steadily through the magnificent scenery, mist and cloud to the beautiful town of Ludlow for a pub lunch, and from there took us home.

CHAPTER 29
Come to sunny Lincolnshire

After getting back home from Wales, we went out and with our usual impulsiveness, bought a caravan. It seemed a good idea; not only would we make a little money after the fairly large initial outlay, but it was a pleasant thing to do - along the same lines as our earlier working holiday project, but without the disruption to the family; a chance for social contact with likeminded folk, which we were short of out here in the wilds; and also a regular stream of playmates for the children. Perhaps the visitors would even want to work - an added benefit. The van was a six-berth, rejected from a Skegness site, not because of unsound condition, but rather for its appearance, and it cost £265. Against the stark fen landscape in February, the caravan stood out like a sore thumb, with its pale blue and cream sides, but with the coat of dark green paint and a screening of huddled trees and hedging that we intended, it would blend into its surroundings.

We had the caravan rewired, fitted with a slot meter and connected to the mains in the house. We then went out in search of second-hand equipment for the hundred and one odds and ends that are needed in a home on wheels - blankets, washing-up bowls, pots and pans, cutlery, and much, much more. It was surprising even to us just how cheaply we did scrape together the necessary. A chemical toilet was already supplied, situated in its own cubicle in the cosiness of the van - no stumbling by torchlight to a tent somewhere out in the open, only to find it has blown over in the fen gales! Rick dug a soakaway to take the effluent from the kitchen sink, but the water would have to be collected from the toolshed tap for the present; we wanted to keep costs to the bare minimum at the beginning. If it proved to be an economical proposition we could lay on mains water later. We took out the tiny calor gas cooker and bought a decent-sized domestic electric one for £10 from an advertisement in the paper. We have now been given a

small refrigerator too, which will be useful for storing milk, meat and so on. We also included a woodburner for those times when it is not quite so warm.

Now all we had to do was collect the bookings. We toyed with the idea of catering for anglers, it being an extremely popular fishing area, but decided just to appeal to self-sufficiency enthusiasts, even if this meant not having the van as full as we might have otherwise. It was with considerable apprehension that we awaited our very first guests that Whitsun bank holiday weekend - I do not know if the feeling was mutual when Peggy and Eddy saw what they had let themselves in for!

They arrived with bicycles and walking boots and boundless energy. They were to prove ideal visitors - very interesting, and both ardent faith healers. There was much we wanted to ask them about their beliefs, and they revelled in our way of life, wanting to participate and be part of it. We really showed them what life is like in the country - none of the slow leisurely pace that dreams are made of here! Due to circumstances beyond our control, that week was one of our busiest on record - Blossom calved, and we bought in two calves, one of which had to be treated for scours. Cello decided to reveal to one and all her headstrong character, and a good hour was spent pursuing her over strawberries, blackberries, over fencing, round the paddock, and across the vegetable garden. One lunchtime two calves escaped from the local doctor's field, which we were using for grazing at the time. The bees played up too, swarming madly. The sun shone day after day, and so the sweet williams ripened behind us as we gathered them for sale. That week we pulled 930 bunches, which we could not have achieved without Peggy and Eddy's help. Shearing also fell in the middle of the chaos, and then, as a parting gesture, Blossom went down with milk fever the morning they were due to leave - but it was all in a week's work!

We certainly had a varied selection of people in the caravan. We came to know what they would be like by the sort of car they owned, rather like a man looking like his dog! They came from all over, from London and the cities, even from as far as Scotland. Some found our way of life a bit spartan, commenting that this was self-sufficiency at its hardest. There was a couple for whom everything was wrong, the list of jobs almost turning Rick into a full-time odd job man. Then there was Pat and John, who were the complete opposite. They had wet weather for the length of their stay, but not only did they love all of it but returned again three weeks later for more punishment. There was also a gentleman with a brand new car, who parked it in the roadside rather than in our muddy yard. Then there was Cathy and Bill, who arrived from the station on bicycles, their young children, Anna and Jessica, on seats behind with the small amount of belongings they deemed necessary for the week. They saw the countryside in a twenty mile radius, Skegness, Tattershall Castle, and they still had the energy to help me lift potatoes on the final day. Bill was a jazz pianist and we spent the last evening enjoying his playing on Pam and John's piano.

The children, mostly between toddler age and seven, provided a string of friends for our four. Jade considered himself a bit above them, until Penny arrived with her two very attractive daughters. Once the ice was cracked (the older the child, the longer this took), Jade spent the remainder of their holiday assisting in a spider and caterpillar hunt, trying shyly to impress. The younger girl was an insect fanatic, possessing an extensive collection in her bedroom and a knowledge of them, complete with Latin names, that would have been impressive coming from someone three times her age.

When the season had ended and all the holiday-makers had gone back to their homes and their jobs for winter, in spite of the success of the enterprise, we heaved a sigh of relief to be given back our tranquillity. But it is always with eagerness that we look forward to the arrival of the first caravan guests of the next year.

CHAPTER 30

Taking stock

The day we finally took what remained of our belongings in the Landrover and followed the removal van to Stickford, we also crammed in four thousand strawberry plants. We allocated the acre immediately behind the house and outbuildings to the soft fruit block and gradually, over the following two weeks, along with the blackberries and raspberries, grain, grass, roots, kale and vegetable seeds, these strawberry plants were carefully set into the earth.

The area is predominantly one of large farms. There are not so many smallholders, most householders just growing flowers and maybe some vegetables to supply the house. Several 'pick your own' soft fruit establishments have set up to supply a demand; an adventure that would have surely failed in the Wisbech area, where strawberries sold at rock-bottom prices. So we now had the local public to cater for, as far as strawberries were concerned. When the strawberry season came around, although only three months after planting, there was fruit to be picked. One afternoon, we filled what punnets we could with the good quality red strawberries and, together with bunches of new carrots, sold them door to door at thirty pence a pound around Stickford. They were very popular and that evening one of our customers telephoned to buy the remainder of the crop, about 100lb in all.

We had fed the plants one bag of granule fertiliser that March - not yet having a rotted supply of organic manure for a mulch - but omitted to dress them the following spring. Even so, by the time May was out the foliage was green and massive, they were as high from the ground to the tops as the length from elbow to fingertips. No one had seen any so tall before! Not being in a strawberry growing area, we thought it safe not to use pesticides against red spider. The crop ripened late, shielded by the mass of leaf, but when the green changed to white, to pink, and then became red, what a quantity of fruit there was. Rick and I, so naive, had estimated that we alone could pull half an acre, judging by our experience at Ivy Cottage. The sun shone down on us, as we hurried and picked to the best of our ability. Of the forty-eight rows, we were only completing one row each per day. Mr Green, who owned a garden centre on the A16, collected 36 lb a day from us; the remainder we delivered to a very successful wholesaler eight miles away - he was paying us top price at twenty-eight pence a pound for our top quality fruit, with promises to take all we had.

We continued at this frantic pace a day or two longer, until we admitted it was hopeless. An evening was spent telephoning friends, acquaintances, acquaintances of friends and asking around the neighbourhood for pickers. Our faithful band of helpers were loyal to us and laboured hard for their piece-rate wages. The rate was seven pence a pound to pull straight into punnets, and as an incentive only to include good fruit, we encouraged our staff to take undersized, slug damaged or over-ripe fruit home for jamming. Rick found most of his time went on keeping them supplied with the cardboard boxes he had to assemble.

But still we were making no impression on the bounteous crop; a good picker finished his or her row that day, but the following morning, there would be part rows to be gone over again. So, with the sweltering heat still bringing on the strawberries, the pickers not keeping up with them, and the sale of them now limited, something more had to be done at once; the fruit would not hold much longer. Petrol prices made delivering small quantities to local retailers uneconomical, and to London out of the question, especially as a perishable commodity needed transporting daily. We had contacted Organic Farmers and Growers, because after all, as our produce was organic, it would be nice to sell it as such, and to maybe receive the twenty-five per cent premium above market price it deserved, but we had no success. We had already examined the possibility of sending to Manchester and other large markets, but the cost of transport and packaging was prohibitive - on an estimated 5,000 lb of fruit, the above charges exceeded £400 excluding labour. The only alternative seemed to be 'pick your own', which we had purposely avoided, imagining it would cause extensive crop damage. So Rick knocked together temporary signs that evening, which he erected on the main road early next morning.

The first people, arriving at 9 a.m., were a couple *en route* to another 'pick your own' establishment six miles further on; then a

couple from Sutton-on-Sea, who bought vegetables too, begged a bag of manure and were to return again and again. After this, more and more folk came for strawberries, ready-picked and 'pick your own', and for vegetables, and we were much too busy running between our gang of pickers and the customers to remember them all. We found we had to grab a bite to eat when we could, as invariably someone would turn up as we sat down to dinner, tea and even our bedtime drink. Once the piece-workers were paid and the trays delivered to the wholesaler, we could sit back and take the money – a novel experience!

The initial day of 'pick your own' had obviously been a huge success. Eight o'clock the following morning came round and our band of workers very soon arrived, chatting as they took the next row along, ten bottoms in the air showing that pulling was in progress. The 'pick your own' section, marked off by fertiliser bags, was slower than at the weekend, but there were still some customers to attend to and vegetables to fetch.

Then the telephone rang. It was the wholesaler cancelling today's fruit and any more until further notice. Immediately we stopped all operations and paid everyone off, but we had thirty-six twelve pound trays of strawberries on our hands. One of our workers was sympathetic and out of the goodness of his heart bought a handful of punnets and went right into Boston to sell them. Rick rang here, there and everywhere and at last contacted a shopkeeper in Skegness who normally passed near this way going to our wholesaler, and he was prepared to inspect them. He complained it was late in the day for buying perishables and so we had to be content with a rock-bottom price to clear them. However he placed a daily order with us for 108 lb, sometimes more. He soon, together with our booming 'pick your own' trade, cleared us out of all vegetables, carrots, spring onions, beetroots and lettuces. Rick and I were able to meet our daily quota of twelve trays ourselves, while the 'pick your own' side continued to flourish. We moved the flags over regularly to give the fruit a chance to ripen and to ensure there was fruit to pick, folding the pickers over the strawberry block as one folds sheep over kale.

The success of this enterprise must have gone to our heads. We planned our 1982 season for months, putting down a further half acre of maidens, from which we removed the fruit buds the first season to allow all the energy to be put into plant growth. We planted up the vegetable area solely to coincide with the 'pick your own' market, presuming there would be some raspberries and blackberries too. People loved having their vegetables as fresh as possible, even choosing which size lettuce to cut. We were charging no more for our organic husbandry though. Thus we planted a quarter of an acre of spring onions and beetroot, an eighth of an acre of carrots, a quarter of an acre with runner beans and some radish and lettuce. Ideas for this 'strawberry fields' venture as we called it, were our talking point all along, as we drilled, hoed, singled out and weeded the vegetables on hands and knees, and the strawberries likewise. With the latter, no sooner had we reached the furthest row, than those nearest were swamped with weeds.

An official from the Weights and Measures Department arrived one day to check our scales. He took out of his suitcase a set of weights and became engrossed in his work. Our scales were found to be accurate only on small quantities; by the time the dial had swung round to 20lb, they were out by several ounces. This had to be sorted out prior to the commencement of the season and we had to order a new set of scales at a cost of £135 - money we could ill afford.

The extensive dry spell of that time, was finally broken by a period of frequent showers, often developing into unremitting downpours. All of which had an extremely adverse effect on the strawberry bed - the rows became denser with both weed and foliage growth, underneath which formed trusses lay on a liberal mulch of chicken manure mixed with sawdust. Rain water trickled along the leaf stalks and onto the trusses. Air circulation amongst the undergrowth was non-existent, and in this dampness the trusses began to rot. Nearly all the early flowers had been caught by a severe late frost, any that had escaped and developed into a strawberry, now appeared as a furry, grey fungus. The botrytis became more and more widespread. As it had occurred so early, trusses were just disintegrating, rather than the usual case of rotting fruit. It gradually became apparent that there would hardly be a crop at all.

We notified ADAS at the first sign of the disease, but a fungicide is used in a conscientious programme as a preventative, rather than a cure. After other growers in the area had been advertising their produce for a fortnight, we reckoned there were a small number to be salvaged. For weeks, previously satisfied customers had been telephoning and calling in to enquire when our strawberries would be ready, so we began to collect orders for ready-picked fruit, probably only of jamming quality. With much searching, we could only unearth one pound of strawberries per fifty yard row, so obviously we could not open the field to 'pick your own'. All the time and effort involved probably was not worth it, but I fought to the end, not prepared to let the chickens (which we still allowed to range free this year on the devastated area) gobble everything.

We now entered a time of deep thinking. Had we sprayed with a fungicide, we might have saved forty per cent, the figure given by the Ministry for commercial growers in the area, whereas our 130lb strawberries (as compared with 6,300lb in the previous year) showed a salvage rate of little over two per cent. For eight years we had grown organically, giving tender care to the life force of the soil. But for all our efforts, we could not secure an organic market for our produce and to survive disasters like this, the higher reward is essential. The weather also had produced an incredible weed problem. We had battled for fourteen hours a day, but for all our fighting we had lost. The kale stood feet below the thistles and willow-weed; and butterflies fluttered amongst them. Spring onions were lost in the shepherd's purse and woody nightshade. Around us stood the immaculate fields, probably only surveyed once a week, of large farms. Our tiny plot that swallowed all our working hours looked semi-derelict. We were no longer earning money from 'pick your own' sales, we had no wholesaler, and the retailer in Skegness was not prepared to come out of his way without the strawberry incentive. The owner of three wholefood stores in the main local towns bought six bunches of onions, carrots, beetroot and radish on a Tuesday evening, at the usual wholesale price, and by Saturday his family had to eat most of the wilting vegetables themselves. An advertisement in the local health food shop for our organic soft fruit and vegetables brought no response whatsoever.

All around appeared depressed and neglected, and we felt that way too. It seemed sheer folly to sacrifice our livelihood for a principle and one which no one around here valued. It was obvious that we would now use a fungicide. Our thoughts on insecticides were also easily revised – if Blossom goes down with milk fever we call in the vet to treat her chemically, and so it must be with our crops. If disease attacks – carrot fly, onion fly and so on – then we must treat it as best we can, for no one will buy bunches of carrots with yellowing leaves at a premium organic price. But herbicides were a harder decision. Without the constant battle against weeds, we could crop far more intensively. Unable to secure a higher price for quality or organic produce, the alternative was to make more money by growing more produce, but we could not tell what effect our actions would have on all the teeming millions of tiny creatures within the earth. Thus the gloom that prevailed was a result of one year's loss of main income, estimated well into four figures, a year's set-back as nature took its course, just when we thought we were building up a prosperous 'pick your own' trade. But the low spirits were more as a result of abandoning all we had believed so right and true in our eight years of organic husbandry. I must emphasise that our decision to use certain chemicals, to reach a compromise between the two extremes, was only reached after much soul-searching. Our decision to use one or two carefully selected chemicals was not for financial gain alone. We do not seek wealth and possessions, we only want a reasonable income with which to feed and clothe our family and pay our way with complete independence. At Fir Tree Cottage we have been constantly under pressure to get various things achieved. This pressure and its associated stress is what we moved to the country to leave behind. We could never have continued at this pace into our old age. We now have a holiday occasionally; there must be more beyond our four dykes and we want to see it. With an easier pace we can really come to enjoy the everyday things in life again – feeding the calves at leisure, watching the hens peck up their scattered grain, collecting greenstuff for the rabbits – all of which at present have become just a long list of chores to be done each day.

Thus it was with a completely clear mind that we finally called in the man from the Ministry to advise us on the use of the safer sprays, bearing in mind our organic tendencies and our continued application of farmyard manure. I think the poor chap was completely taken aback at first at the overrun state of affairs, but we found him to be helpful. He explained that feeding the strawberries with manure had been the main contributing factor to the vast amount of top growth. Our soil type (peaty loam) was already extremely fertile and the addition of more nitrogen only produced leaf. He explained that burnt straw sprinkled on the rows as potash would be quite sufficient – the practice of burning off the straw down the rows preceding cropping not only destroyed the harmful strawberry pests, it also had a fertilising effect. The official systematically went through our crops, advising us as was appropriatc. Hc listcd thc numcrous weeds to be exterminated, and would report back on the remedy; the list included willow-weed, nettle, thistle, redshank, chickweed, pineapple mayweed, sow thistle, columbine, black bindweed, coltsfoot, goosegrass, willow-herb (Rosebay), dock, woody-nightshade, creeping butter-cup and shepherd's purse. I still wanted the nicer ones amongst our vegetables, the heartsease, forget-me-not, speedwell and the scent of the wild mint. He saw no reason to prevent us earning a fair income by cropping more intensively. He also stated that in all of his thirty-five years with the Ministry of Agriculture he had never known a strawberry season like it; we hope another such will not be bestowed upon us in our lifetime, but if it should be, we will be better prepared.

CHAPTER 31

But there is nothing else we would rather do

I hope I have shown the down-to-earth realities of smallholding life as it really is, rather than the popular idealised version. There is a large failure rate of people who descend on the country, blinkered by their high ideals and theories. After a couple of years struggling, often on the dole, with a fair share of troubles, vets' bills and so on, which seem out of all proportion, the cons seem to outweigh the pros and they pack up and go back. Many is the time now we have picked over the sorry leavings. They arrive with such fixed opinions of their own, that they immediately alienate themselves from the men of the soil they seek to imitate; men born into the farming community, who have worked the land for as long as they can remember and could teach them so much, if only they would stop to listen a moment. They are unable to see that what they are seeking cannot be learned from text books, only by experience.

Many of our original self-sufficiency friends have changed direction. Those who have not finally abandoned it, have certainly mellowed somewhat; the idealists have aged; the fanatics have slowed down, including us. For us the awakening came this summer. We had become too serious over our way of life, and we are no longer 'wheelbarrow farmers'. The countryside should be a place in which to play as well as work.

As a result of our new philosophy that life should be fun we spent a week in Devon in November, partly with the intention of spying out the land. The bad year had made us think of moving, but as soon as we got there, we realised that the fens were home – a nice feeling. Moreover, we would be hard put to make a living there in the only way we know how (market gardening) and the high water table would also have depressed us. We travelled round to see the sights in the rain, and it served to get everything back in perspective again.

We came back renewed to the fens, where although the geo-

graphical layout might be a bit bleak, it holds a beauty of its own and the sunsets are amazing. The scattered holdings in traditional style offer as much interest and stimulation as we need on country strolls and the landscape of the sky makes up for it all.

We returned with plans for the land (more grass, so that what arable land we keep is well kept) and for the house (a porch, an extra bedroom). Rick intends to develop his woodturning, producing pots, bowls and so on that he can sell in the strawberry season. I am going to experiment with clay, just producing primitive objects with a useful purpose.

One must always go forward. Everything is always changing – the aims we started out with have altered, ideas have even changed direction in the seven months it has taken to write this down. Nothing stands still. We are still scatty to this day! Who else would waste an hour chasing two escaped rabbits that were rapidly ruining our winter supply of brassicas – at twenty pence a cabbage it puts meat at a high price indeed per pound! But there is nothing else we would rather do.

APPENDIX

Vegetable dyeing

DYESTUFF	MORDANT	COLOUR
acorns	alum	grey
apple bark	alum	beige
beetroot	alum	pinks
blackberries	alum	purple
shoots	alum	yellow
	iron	grey
	copper sulphate	green
carrot leaves	alum	green
chives		
seedpods	alum	pale grey
stems	alum	pale yellow
clay (terracotta)	alum	cream (and softens the wool)
coltsfoot	alum	green
copper sulphate	none	varying shades of green depending on length of contact
cutch	alum	pinks (specially recommended)
daffodil	iron.	purple
	alum	lemon
	copper sulphate	lime
	chrome	sage green
dandelion heads	alum	pale yellow
roots	alum	beige
dock leaves	alum	sage green
elderberries	alum	pinks and purples (depending on quantity of berries used)
	chrome	bronze/brown

DYESTUFF	MORDANT	COLOUR
leaves	alum	yellow
golden rod	alum	yellow
gorse	alum	lemon
henna	alum	orange - after first dyebath brown - after second dyebath
hogweed	alum	sage green (specially recommended - strong dye)
ivy	alum	beige (disappointing)
kale	alum	pale beige
laburnum bark	alum	bronze/brown
mare's tail	alum	pale beige
marigold petals	alum chrome copper sulphate	sunshine yellow bronze/brown sage green
meadow-sweet		
roots	alum	pink/beige
leaves and flowering tops	alum	sage green
nettles	alum iron copper sulphate	pale green grey green
oak bark	none	brown
onion skins	alum copper sulphate	yellow dark green
pea pods	alum	off-white
plantain	alum	off-white
poppy petals	alum	pale mauve
privet	alum	yellow
rue	alum	green
sage flowers	alum	pale green
red sandstone	alum	salmon pink
sedge grass	alum	mustard (but fades)
shallots	none	beige

DYESTUFF	MORDANT	COLOUR
	alum copper sulphate	bright yellow green
silver birch bark leaves	alum alum	brown bright yellow
sorrel	alum	green
sow thistle	alum	pale yellow
spinach	alum	mustard yellow
sweet williams	alum	metallic green
tansy	alum	bright yellow
tea	alum	brown
teasles	alum	nondescript
thyme	alum	pale green
tomato vines fresh dried	alum alum	pale green pale beige
turmeric	alum	bright orange/yellow
walnut haulms	alum	deep chocolate brown